Nuclear Nonproliferation: The Spent Fuel Problem

(Pergamon Policy Studies—32)

Pergamon Policy Studies on Energy and Environment

Cappon—*Health and the Environment*
Murphy—*Energy and the Environmental Balance*
Goodman/Love—*Geothermal Energy Projects*
Constans—*Marine Sources of Energy*
De Volpi—*Proliferation, Plutonium and Policy*

Related Titles

Fazzolare/Smith—*Changing Energy Use Futures*
Myers—*The Sinking Ark*
Wenk—*Margins For Survival*
McVeigh—*Sun Power*

PERGAMON POLICY STUDIES ON ENERGY AND ENVIRONMENT

Nuclear Nonproliferation: The Spent Fuel Problem

Edited by
Frederick C. Williams
David A. Deese

Published in cooperation with the
Center for Science and International Affairs,
Harvard University

Pergamon Press
NEW YORK • OXFORD • TORONTO • SYDNEY • FRANKFURT • PARIS

Pergamon Press Offices:

U.S.A.	Pergamon Press Inc., Maxwell House, Fairview Park, Elmsford, New York 10523, U.S.A.
U.K.	Pergamon Press Ltd., Headington Hill Hall, Oxford OX3 0BW, England
CANADA	Pergamon of Canada Ltd., 150 Consumers Road, Willowdale, Ontario M2J 1P9, Canada
AUSTRALIA	Pergamon Press (Aust) Pty. Ltd., P O Box 544, Potts Point, NSW 2011, Australia
FRANCE	Pergamon Press SARL, 24 rue des Ecoles, 75240 Paris, Cedex 05, France
FEDERAL REPUBLIC OF GERMANY	Pergamon Press GmbH, 6242 Kronberg/Taunus, Pferdstrasse 1, Federal Republic of Germany

Copyright © 1979 Pergamon Press Inc.

Library of Congress Cataloging in Publication Data
Harvard University. Nuclear Nonproliferation Study Group.
Nuclear nonproliferation.

(Pergamon policy studies)
Includes index.
1. Atomic power plants—Waste disposal. 2. Spent reactor fuels. 3. Nuclear nonproliferation.
I. Astiz, Carlos Alberto. II. Williams, Frederick C., 1938— III. Deese, David A. IV. Title.
TD899.A8H37 1979 614.7'6 79-4535
ISBN 0-08-023887-4

All Rights reserved. No part of this publication may be reproduced, stored in a retrieval system or transmitted in any form or by any means electronic, electrostatic, magnetic tape, mechanical, photocopying, recording or otherwise, without permission in writing from the publishers.

Printed in the United States of America

363.728
N964

To our families,

with gratitude for their support,

encouragement, and understanding

Contents

Foreword

Each year the world's nuclear power reactors produce enough plutonium to make well over a thousand nuclear weapons. As long as this plutonium is contained within the spent fuel elements, there is little cause for concern; in this unseparated form it cannot be used to make bombs. However, an incipient worldwide spread of spent fuel reprocessing facilities threatens to make separated plutonium, which is suitable for use in weapons, available to an increasing number of countries. The expansion of national stockpiles of plutonium, and the increased capability to produce such stockpiles rapidly would exacerbate the nuclear proliferation problem perhaps more than any other aspects of the power reactor fuel cycle.

Spent fuel must be stored pending either later reprocessing or permanent disposal. Although spent fuel storage is a relatively simple technical task, the economic, political, and institutional dimensions surrounding it are far more complex.

For some nations, there is a ready solution: Store the spent fuel in facilities at the reactor sites or develop national facilities elsewhere. For others, it is not so easy. Anti-nuclear activists point to interim storage of spent fuel as evidence of the absence of an acceptable solution to the problem of permanent disposal of nuclear wastes; they oppose prospective storage sites on the basis of safety or adverse environmental effects; and they question the adequacy of physical protection against theft or sabotage.

For very different reasons, nuclear advocates might join in the opposition to spent fuel storage facilities. Their objections could be expected to focus on the deferral of reprocessing rather than on the storage of the spent fuel. They might cite the economic advantages of recycling the plutonium in the spent fuel to supply existing power reactors, or the technical advantages of reprocessing for waste disposal. They might stress the need for gaining experience with plutonium fuels before the breeder age arrives, or the proliferation dangers inherent in keeping the plutonium in the spent fuel.

Given these potential barriers to the establishment or expansion of national storage facilities, it is essential that alternatives to national facilities be explored. A report on one such exploratory study is the essence of this volume.

A broad spectrum of possible multinational and international arrangements for spent fuel management is covered, ranging from relatively benign international oversight of national facilities to arrangements for bilateral and regional cooperation, and even to the creation of entirely new international institutional mechanisms. Thoughtful consideration is given to the technical, economic, political, and legal aspects of managing spent fuel. Eastern Europe, Western Europe, the Indian Ocean Basin, Asia, the Middle East, and Latin America are of particular concern in this regard.

The study was conducted at the Center for Science and International Affairs at Harvard University, with funding provided primarily by the Ford Foundation and the U.S. Department of State. A working group of fifteen scholars and professionals, representing diversity of academic levels, professions, experience, and nationalities, met biweekly over the course of a year to discuss emerging ideas. They conferred with invited speakers, criticized draft papers, and conducted interviews in the United States and abroad. This project, and the book that grew out of it, was truly a collaborative effort. Frederick Williams was invaluable in guiding this working group vehicle, which had almost as many constructive back-seat drivers as it had participants to its destination; and he and David Deese managed to convert 11 separate written works into a cohesive and important volume. I am confident that readers will share my feeling of indebtedness to all who were involved.

Albert Carnesale

Acknowledgments

We are indebted to the Center for Science and International Affairs at Harvard University for generous financial and staff support, as well as for the rich intellectual environment for research and writing. Our gratitude also extends to The Ford Foundation for its continuing general support of the Center. The chapters in this volume are the products of a full year of intensive interaction among fifteen members of a research working group. The group benefited immeasurably from the leadership and insight of Albert Carnesale, Associate Director of the Center and Professor of Public Policy in the John F. Kennedy School of Government.

We gratefully acknowledge the assistance of many leaders and researchers in governments, universities and industry in the United States, numerous foreign countries and several international organizations. We are particularly indebted to the International Atomic Energy Agency, especially John Colton, and the Australian Embassy in Washington, D.C., for their generous administrative support of our field research efforts. Jeffrey McAllister provided important editorial assistance in the attempt to convert the individual chapters into a coherent and consistent volume.

The authors and the Center are grateful for the general support of a research grant from the U.S. Department of State, which covered most of the meeting, administrative and travel expenses for the working group. The contributions of Deese and Marwah were supported in part by International Relations Fellowships from The Rockefeller Foundation. The views and conclusions contained in this study are those of the authors as individuals and should not be attributed to the U.S. Government, The Center, or the Foundation.

Introduction

With some determination and a handful of plutonium, a large number of countries can now build a nuclear explosive device. As nuclear technology and nuclear engineering experience spread around the world, the capability to construct nuclear weapons diffuses. In large measure, the engineering capabilities follow the growth of the commercial nuclear power industry in smaller countries, where the first appearance of plutonium is in the spent fuel removed from a commercial nuclear power reactor. While there are many avenues to proliferation, the separation of plutonium from reactor fuel is one route that presents a serious danger.

A potentially severe spent fuel storage problem complements this danger in a number of the more advanced countries with larger commercial nuclear power programs: in the United States, rapidly-filling storage pools threaten possible reactor shutdowns in the 1980s; in Austria and Germany, nuclear power programs have been halted permanently or temporarily because of inadequate alternatives for spent fuel disposition.

In his April 7, 1977 policy statement on nuclear energy and the proliferation of nuclear weapons, President Carter fully addressed the first problem but underplayed the second. The announced policy included the indefinite deferral of commercial reprocessing and recycling of plutonium produced in American nuclear power reactors, the deferral of the commercial use of breeder reactors, and the establishment of an international Nuclear Fuel Cycle Evaluation (INFCE) program. This program would seek to develop alternative nuclear fuel cycles and to formulate international measures to assure access to nuclear fuel supplies and spent fuel storage for nations sharing common nonproliferation objectives. However, many countries, including Britain, France, West Germany, Japan, and the Soviet Union, have not followed the American example and are proceeding with plans for commercial reprocessing and commercial use of breeder reactors. Through INFCE, the United States is now suggesting an alternative course of action. It proposes the establishment of international management over the storage of spent reactor fuel, and international cooperation in establishing facilities for the storage of spent fuel. That proposal is examined in this book. While it can be viewed as a response to the problem of weapons proliferation risk, its main thrust is really directed to the reverse side of the nuclear energy coin: the proper management of reactors and their fuel cycle, and the appropriate response to energy planning problems associated with spent fuel disposition have been too long neglected by the Administration's policy.

NUCLEAR WEAPONS

Before examining the political and economic issues involved in the international management of spent fuel, it is helpful to recall the physical processes involved in nuclear weapons and nuclear power reactors.(1) Nuclear explosive devices rely on the physical phenomenon of nuclear fission. When neutrons strike certain unstable isotopes of uranium or plutonium, these atoms split into lighter elements and emit both a large amount of energy and more neutrons. These neutrons, in turn, strike more uranium or plutonium atoms, which split and release more energy and more neutrons. The result is a chain reaction of splitting atoms, with an enormous release of energy. The complete fission of one pound of fissionable material releases energy equivalent to that released from the explosion of about 8,000 tons (8 kilotons) of chemical explosive such as TNT.

An atomic nucleus is fissionable if it can be split when struck by a high energy (fast) neutron. An atom is fissile if it can be split when struck by a high energy or low energy (slow) neutron. Many heavy atoms are fissionable, but only a few are fissile. Fissile atoms include uranium-233 (U-233), uranium-235 (U-235), and plutonium-239 (Pu-239). Since the neutrons emitted from a nuclear fission include both fast and slow neutrons, fissionable atoms alone would not sustain a chain reaction. A bomb requires fissile atoms. Uranium-238 (U-238) and thorium-232 (Th-232) are heavy atoms; they are fissionable but not fissile.

Moreover, a small amount of fissile material will not sustain a chain reaction because too many neutrons escape through the surface and are not available to collide with other nuclei. For a given shape or geometry, the smallest quantity of fissile material required to sustain a chain reaction is called the critical mass. The geometry for which the critical mass is smallest is the sphere. Spheres of about 50 kilograms (kg), 12 kg, and 10 kg are needed for critical masses of U-235, U-233, and Pu-239, respectively.

Six countries have tested nuclear explosives. The United States, the Soviet Union, The United Kingdom, France, and India constructed their first nuclear devices with plutonium. Only China used uranium in its first atomic bomb.

The fissile isotopes – U-235, U-233, and Pu-239 – do not occur abundantly in nature. Natural uranium ore consists of about 99.3

percent U-238, and only 0.7 percent U-235. Neither U-233 nor Pu-239 occur in nature; they are produced in power reactors, U-233 coming from Th-232 and Pu-239 from U-238. It is not necessary to have pure U-235 or Pu-239 in order to construct an atomic bomb. In a uranium weapon, the fissile core consists of highly-enriched uranium with more than 90 percent U-235.

Reactors designed for producing weapons-grade materials can produce relatively pure Pu-239 with little contamination from Pu-240 or other plutonium isotopes. Unlike these production reactors, reactors that are designed to generate electric power produce plutonium mixed with various plutonium isotopes. The plutonium in the spent fuel from a light water reactor may consist of about 60 percent Pu-239, 21 percent Pu-240, and 19 percent heavier isotopes. However, a nation that decided to acquire nuclear weapons, but had no supply of weapons-grade plutonium, could still design and build weapons with reactor-grade plutonium.

NUCLEAR FUEL CYCLE

All commercial power reactors use U-235 as their source of energy. After removal from the mine, uranium ore is milled by mechanical and chemical means to obtain a concentration of between 70 and 80 percent uranium oxide (U_3O_8), or "yellowcake." This concentrated (U_3O_8) then goes to a chemical plant where it is converted to uranium hexafloride (UF_6). However, light water reactors require uranium fuel containing between 2 and 4 percent U-235. Therefore, the UF_6 goes next to an enrichment plant where, by gaseous diffusion, centrifugation, or some other technology, the uranium is enriched from 0.7 percent U-235 to the necessary 2 to 4 percent U-235. After enrichment, the uranium is converted to a form chemically suitable for use as a nuclear reactor fuel, either uranium oxide (UO_2) or uranium carbide (UC).

In a fuel fabrication plant, uranium oxide powder is compacted into small pellets; the pellets are stacked in a thin-walled metal tube or cladding, which is then sealed. Several of these fuel rods are bound together to produce a fuel assembly, or fuel bundle. When recycled plutonium is used with uranium in the reactor fuel, plutonium oxide powder and uranium oxide power are mixed together and compacted into pellets.

During operation of the light water reactor, the processes of neutron capture and beta decay convert U-238 into Pu-239. Subsequent neutron capture by Pu-239 produces Pu-240 and even heavier isotopes of plutonium. In general, the longer the fuel rods stay in the reactor, the heavier are the plutonium isotopes produced, and the less suitable is the plutonium for weapons.

The "back end" of the fuel cycle begins when the spent fuel rods are removed from the core of a nuclear reactor. This spent fuel is extremely radioactive, and must be kept under water in storage pools at the reactor site for at least a few months while the level of radioactivity declines to manageable levels. In this spent fuel, one encounters, for the first time in the light water nuclear fuel cycle, material usable for weapons, namely Pu-239. Because of a lack of alternative storage sites, and for other reasons, spent fuel has accumulated in reactor storage pools in many countries, including the United States.

From the storage facilities, the spent fuel can be sent either directly to a waste disposal repository or to a chemical reprocessing plant. There, Pu-239 and uranium can be extracted for re-use as reactor fuel in light water reactors or breeder reactors, or for weapons purposes; and radioactive wastes can be put in a chemical and physical form suitable for permanent disposal. These highly-radioactive wastes include heavy actinide nuclei and fission products that must be isolated from the environment for thousands of years. The debate over the appropriate means of permanent storage of nuclear wastes is unfinished, but it is clear that the siting and timing of permanent nuclear waste disposal facilities is intimately related with the siting and timing of interim spent fuel facilities.

REPROCESSING AND SPENT FUEL STORAGE

This volume concentrates on the issue of the appropriate management of spent reactor fuel. In the fuel cycle of light water reactors, reprocessing spent fuel to obtain plutonium is the only place in the nuclear fuel cycle where materials appear that are directly usable in the fabrication of nuclear weapons. As part of the Atoms for Peace Program in 1955, the United States made public the basic technology for reprocessing; it is now easily available in the open literature. It has been estimated that it would take three to seven years and cost $10 million to $25 million for a nation to build a small reprocessing plant to extract plutonium for a weapons program.(2) India has already done so, and Argentina is building an "experimental" scale facility.

As yet, plutonium's economic use as recycle fuel in light water reactors has not been demonstrated, nor is its large-scale and widespread use as breeder fuel imminent; commercial breeder introduction is a good twenty years away. Reprocessing as a method of managing spent fuel and providing for the disposition of wastes is therefore premature. The easy availability of plutonium attendant to such a method of waste management on a national basis makes it difficult to control the spread of nuclear weapons.

This volume studies an alternative to immediate reprocessing as a method of managing spent fuel inventories: Spent fuel is placed in interim storage under international oversight and cooperation. Even so modest a proposal is bound to be complicated and controversial. Each nation has its own nuclear ambitions in both weapons and energy; each has its own political objectives and economic realities; and each has its own regional and global security problems. There are technical, legal, and administrative problems inherent in the establishment of international arrangements to manage spent fuel. There are incentives and disincentives to participation in an international storage regime. The following chapters examine these issues in considerable depth.

OVERVIEW

Part I, a single chapter establishes a framework for the rest of the book. It consists of an institutional and political analysis by David A. Deese and Frederick C. Williams. The authors develop a model for international spent fuel storage based on the assumption that the international political system changes only incrementally. A series of modest steps toward the institutionalization of spent fuel storage can form the basis of a compromise of conflicting interests. Deese and Williams also develop a set of siting criteria which form an analytical tool for choosing a spent fuel storage site.

Part II puts forth a region-by-region analysis of the political feasibility of regional or multinational arrangement for storage of spent fuel. The chapter discusses first the evolution of Soviet nuclear and nonproliferation policy in relation to the West, China, and Eastern Europe, and the relevance of that policy to the prospect of Soviet participation in an international spent fuel storage regime. It is concluded that the Soviet spent fuel storage regime is, perhaps, less exemplary than it appears; but Moscow's strong opposition to the proliferation of nuclear weapons suggests the possibility of joint Soviet-American efforts toward the establishment of other regional spent fuel storage regimes. The prospects for a regional spent fuel storage regime for Indian Ocean states are then analyzed. An examination of the political tensions in India, Pakistan, and Iran leads Onkar Marwah to conclude that a regional spent fuel storage facility sited in any of the three countries is improbable.

In Chapter 4, Victoria Johnson and Carlos Astiz survey nuclear energy in Latin America. Mexico, Argentina, and Brazil have substantial nuclear programs without imminent spent fuel storage programs. After a study of the political relationships in Latin America, the authors conclude that a spent fuel storage facility in this region is possible.

In Chapter 5, Richard Broinowski studies the nuclear power programs and political concerns of the nations of Asia, the Middle East, and Australia. China would not participate in an international spent fuel storage regime; nor is Japan likely to participate because of its plans for reprocessing and breeder reactors. However, other Asian countries might participate; following criteria proposed by Deese and Williams in Chapter 1, Broinowski rates the suitability of Asian countries as hosts for a regional spent fuel storage facility. He also observes that political differences make it unlikely that a spent fuel regime could be established in the Middle East.

In Chapter 6, Robert Gallucci considers the political and economic issues involved in spent fuel storage in Western Europe, with particular emphasis on Britain, France, and West Germany. In Western Europe, a crucial consideration is easy accessibility to, and retrieval of, the plutonium content of the spent reactor fuel that a country has deposited in a multinational storage facility.

Part III concerns the feasibility of international spent fuel management. In Chapter 7, Marvin Miller surveys different storage modes, and how accident, sabotage, and transportation considerations differentiate between them. He explains how the radioactivity of spent reactor fuel serves as a barrier to diversion and reprocessing, and how this barrier decays as a function of time.

In Chapter 8, Boyce Greer and Mark Dalzell give an economic analysis of the costs of multinational spent fuel storage, and conclude that it is economically advantageous for a country with a small nuclear program to participate in a multinational storage regime, even if its spent fuel must be transported a great distance.

In Chapter 9, Daniel Poneman surveys the various political incentives and disincentives to participation in an international spent fuel storage facility. Incentives include assurances of future supplies of enriched uranium fuel and the avoidance of domestic political opposition to nuclear waste disposal. The major disincentive is the loss of autonomy.

Finally, Part IV examines the broader political context of nuclear energy. In Chapter 10, Dorothy Zinberg traces the history of public opinion regarding nuclear energy and the role of American scientists in shaping that opinion. She documents the growing importance of public participation in nuclear decision-making and, in particular, highlights the increasingly strong public reactions toward nuclear waste disposal. She notes that public skepticism of governmental pronouncements on nuclear energy stems from past secrecy and deception surrounding nuclear programs. Scientists and engineers can make a significant contribution by dispelling this atmosphere of suspicion.

In Chapter 11, Tariq Osman Hyder considers the proposal for international management of spent fuel from the point of view of North-South politics. He concludes that participation by some important developing countries is unlikely unless more attention is paid to the problem of discrimination, and unless a lessening of vertical proliferation can serve as an inducement to halt horizontal proliferation.

NOTES

(1) For the following discussion of nuclear weapons and the nuclear fuel cycle, I have relied heavily on Ted Greenwood, George W. Rathjens, and Jack Ruina, Nuclear Power and Weapons Proliferation, Adelphi Paper No. 130, International Institute for Strategic Studies, London, 1977.

(2) Greenwood, Rathjens, and Ruina, ibid.

I

General Framework

1 Legal, Institutional and Political Aspects of Managing Spent Fuel Internationally

David Deese
Frederick Williams

INTRODUCTION

In a few decades at most, the technology of nuclear waste disposal will be sufficiently developed to allow the world's nations to agree on whether they should jettison spent nuclear fuel directly and permanently, or reprocess it for reuse. In the meantime, some nations are aggressively pursuing reprocessing, while others are cautiously deferring or rejecting it. Even the most optimistic advocates of reprocessing believe that inventories of unreprocessed spent fuel are accumulating beyond the level that reprocessing plants will be able to handle.

Environmentalists, industrialists, and government officials in most nations with nuclear power programs agree that the accumulation of large quantities of spent fuel on nuclear reactor sites presents a pressing problem. The environmentalists fear that interim storage facilities at reactor sites will become longer-term dumping grounds, thus removing the pressure to solve the waste disposal problem. Nuclear utilities have trouble running existing power stations where spent fuel storage pools are reaching capacity; and it is difficult to obtain licenses for new stations because no one knows how to dispose of future spent fuel. Government officials cannot confidently include nuclear power in future energy plans because the confusion surrounding spent fuel disposition impedes new construction. And all three groups worry that large concentrations of spent fuel will attract terrorist attacks.

These spent fuel management problems are most severe in countries that have many nuclear power reactors. Even so, India already has a spent fuel storage problem and soon other countries with small nuclear programs will share this difficulty. For most nuclear countries, disposition of spent reactor fuel will become an important national and international issue in the 1980s and 1990s. Some institution – private, national, or international – will have to absorb substantial amounts of spent fuel without overburdening national reactor programs.

This accumulation of spent reactor fuel presents yet another problem. Evidence from the United States, Britain, France, West Germany, Sweden, Japan, and elsewhere demonstrates that, unless other reasonable options are available, spent fuel accumulation, in and of itself, can be a powerful driving force toward national reprocessing. To have large quantities of spent fuel accumulating worldwide, especially in countries without immediate plans for overseas reprocessing or direct disposal, will seriously complicate international efforts to check the spread of nuclear weapons.

If future reprocessing for all countries occurs exclusively in the Soviet Union, the United States, France, Britain, West Germany, and Japan, part of the reason for centralized spent fuel storage evaporates. Reprocessing has already begun in Argentina, Belgium (Eurochemic), Taiwan, India, Italy, and Spain, and is firmly planned in several other nations, including Brazil and Pakistan. National reprocessing plants, with the accompanying threat of atomic weapons fabrication, may soon be dispersed worldwide. If the countries with spent nuclear fuel could develop an internationally controlled system for handling their growing wastes, nuclear energy could have the raw material it needs to expand without stimulating the further spread of plutonium reprocessing capabilities.

Four general categories of countries emerge on the basis of their plans for national reprocessing:(1) 1) The United States, the Soviet Union, France, Britain, West Germany, and Japan are primary nuclear reprocessing countries and countries planning to store foreign spent fuel; 2) Argentina, Brazil, Belgium, India, Italy, Pakistan, Spain and Taiwan are countries with small reprocessing facilities or with plans to construct reprocessing plants; 3) Countries such as Iran and Mexico have no firm present plans, but many decide to reprocess in the future; and 4) Countries such as Austria and Switzerland are unlikely to reprocess in the foreseeable future.

Even countries that build their own reprocessing plants may find economic, political, and institutional advantages in joining a centralized spent fuel storage arrangement. They may also find that it increases regional security. Most countries, however, will participate only if guaranteed that they can retrieve and reprocess the fuel at any time.

Countries that temporarily relinquish their control over spent fuel to an international facility may make demands. They may request additional military assistance to compensate for the atomic weapons they have forsaken, or they may ask for faster reductions in the arsenals of the nuclear weapons states. Certainly if international spent fuel storage is to attract nations – for example, Argentina, Brazil, India, and Pakistan – that already possess all or most of the facilities needed to build nuclear weapons, the prestige of nuclear weapons must be decreased. Some specific areas for progress in this direction include achieving a comprehensive nuclear weapons test ban, reaching agreement in ongoing strategic arms limitation talks and commitments to go further in the future, ceasing production of all new nuclear weapons-grade materials, and opening all nuclear facilities – except nuclear weapons state military production plants – to inspection by the International Atomic Energy Agency (IAEA) in Vienna, Austria.

PRELIMINARY QUESTIONS

The feasibility of international spent fuel management hinges upon four principal issues.

Economic Feasibility

If a new facility for fuel storage is not built on the site of an existing nuclear power or reprocessing plant, getting spent fuel to it will require an extra transport leg.

Countries not planning to reprocess can use an international spent fuel storage facility without concern over later transport to a reprocessing plant. Other countries might be reluctant to bear the substantial safety and security risks and the economic costs of an additional transport leg. Despite the significant costs and risks associated with transport, however, it may well be less risky and costly to ship spent fuel to an international storage site than to store it at the reactor and ship it abroad later for reprocessing.

Analyses of regional storage facility concepts for international facilities serving several countries (see Greer and Dalzell) indicate that storing spent fuel in larger, central facilities may cost significantly less than storage in smaller ones. Economies of scale are most pronounced in capital costs, although they also affect operating costs. Annual storage costs per unit of fuel stored decrease sooner for larger facilities, but in all facilities the annual storage costs decrease sharply, evening out within about five years. Given the recurrent problems of full-scale commercial reprocessing of high burn-up light water reactor fuel, and the growing accumulation of spent fuel on reactor sites, there may be an increasing number of countries willing to bear reasonable transport and storage costs for an interim international storage facility.

Storage Facilities

French, British, and German plans to reprocess foreign fuels commercially may be threatened by a lack of storage facilities. The governments in these countries maintain an interest in reprocessing, both as a commercial venture for foreign contracts and as the only practical means for making the breeder reactor a major energy source. As of 1978, the profitability of commercial reprocessing remains highly questionable. Reprocessing countries may continue to offer their services to foreign customers, but their ability to fulfill existing contracts and the willingness of potential customers to commit themselves to the high costs and uncertainty of a new and complex technology remain highly uncertain.

Even if the reprocessors solve all their problems, there will still be more spent fuel than they can handle through the 1990s. Centralized spent fuel storage facilities should not prejudice later use of commercial reprocessing services.

Finally, there do not seem to be any significant economic losses in waiting to reprocess until it becomes necessary in order to meet breeder reactor development deadlines. Under most circumstances, this would not require full-scale reprocessing operations until approximately five years prior to the start-up of a commercial breeder reactor.

Proliferation Assessment

While it may be necessary to guarantee the right of eventual reprocessing to countries using an international spent fuel storage facility, multinational reprocessing ventures may increase transfer of nuclear technology and legitimize reprocessing worldwide. These proliferation disadvantages must be balanced against the proliferation advantages of replacing individual national reprocessing with internationally controlled plants. Deciding whether national or international reprocessing will retard proliferation most effectively will require very careful study. The proliferation hazards of accumulating plutonium at international reprocessing plants also deserves more study.

Final Disposition

To attract members and hosts for a spent fuel storage facility, the arrangements for final disposition of spent fuel – either return, reprocessing, or disposal – will have to be clearly established beforehand.

Countries could initially commit themselves to choosing from among the three options and then settle on a specific one later. Countries using the same facility could choose different final spent fuel management methods, and different regions could specialize in different approaches. Even if methods and sites for final fuel disposal are best left undetermined, there must at least be specific agreement on options, responsibilities, and a sequence of actions to cover all waste management eventualities.

The chance to get rid of radioactive waste – regardless of reprocessing – may well induce countries to cooperate in the international management of spent fuel. At the least, establishing a strong waste management incentive will require finding a site for spent fuel storage where some system of ultimate disposal, either locally or elsewhere, has been worked out; and an agreement will have to be reached among participants on waste disposal options, timing, and responsibilities. The simple removal of spent fuel from local storage pools could encourage some countries to participate. Having an out-of-country repository for nuclear wastes – "out of sight, out of mind" – would remove an important obstacle to nuclear energy development. And countries not facing this obstacle in the 1970s can count on encountering it in the 1980s.

The international spent fuel facility should be set up to make storage as attractive as possible to member countries. It should not be encumbered by intricate procedures for waste management and spent fuel return that countries might find unacceptable. Fuel return provisions will be necessary to provide potential participants with an adequate level of incentives; some waste management and spent fuel return provisions could induce potential participant countries to store their fuel and meet nuclear reactor licensing requirements and could ensure potential host countries that they would not necessarily be saddled with local waste disposal. These arrangements should be kept flexible and simple; and no waste disposal arrangements should preclude eventual commercial reprocessing.

GUIDING PRINCIPLES

Internalization

The institutions responsible for controlling nuclear energy first evolved as agencies of national governments, but the vigorous export of nuclear reactors and reactor technology in the last decade has made nuclear energy a *de facto* international concern. The somewhat haphazard spread of nuclear energy has brought in its wake both a renewed awareness of the dangers of nuclear weapons proliferation and a realization that solving the proliferation problem will require new international arrangements.

Deferral of Regional Reprocessing

The best developed proposal for internationalizing nuclear fuel management is the regional fuel cycle center (RFCC) that would incorporate storage, reprocessing, and disposal in one internationally controlled facility. It would, in the words of the IAEA, "minimize the number of plants where nuclear explosives could be refined."(2) Yet this concept has foundered on a lack of decisive economic merit and continuing controversy over technology transfer.

Some economic projections say regional reprocessing would make a small net profit, while others say it would lose money; but none has offered a compelling economic argument for reprocessing.(3) More important, internationalization of reprocessing could transfer weapons-relevant technology to technically trained nationals of countries desiring to build nuclear weapons. Eliminating the technology transfer problem might result in inherently discriminatory, and thus unacceptable, treatment of member countries. Without a practical solution to this dilemma or a decisive economic incentive, reprocessing in RFCCs is unlikely to go forward. Some more modest form of international nuclear cooperation must be found to lay the groundwork for more extensive future cooperation, and especially for joint reprocessing of breeder fuel should the breeder attain commercial success.

The Virtue of Modest Steps

Countries make use of nuclear energy through a complex institutional system with political, technical, legal, and economic components. Containing nuclear weapons proliferation will require altering, perhaps thoroughly reconstructing, that institutional system. However, the system is so complex, and its components so closely meshed, that it can easily defy attempts to remold it. Trying to restructure the entire system at once to counter proliferation might cause such discontinuity that antiproliferation goals would be weakened instead of strengthened. For example, if nuclear supplier countries put restrictions on the end use of uranium and uranium enrichment equipment to counter proliferation, the purchasing countries would probably strive that much harder to achieve national fuel cycle independence. In the long run, international control of plutonium would be diminished.

Thus, steps to modify the present international institutions regulating nuclear energy should be modest and carefully crafted. The nuclear energy system cannot absorb any "great leaps forward."

Spent Fuel Storage

Constructing international facilities for spent fuel storage alone, without reprocessing, fits perfectly into this incremental approach. Because reprocessing is stymied for the time being, the need for additional storage capacity is apparent: the expanded Windscale facility in Britain, even if it meets its construction schedule, will not be ready for commercial-scale reprocessing until at least 1987; the German reprocessing plant at Gorleben is still in the planning stage and, even with full-speed development, there will be insufficient reprocessing capacity for German spent fuel for a long time and a consequent need for large-scale storage(4); nor can the French facility at Cap La Hague be counted on to handle remaining European or would needs; the Carter Administration opposes reprocessing and is as reluctant to accept large amounts of foreign fuel as Third World countries are to spend it to a superpower. The world obviously needs more spent fuel storage capacity.

New international storage facilities will not only drain off the spent fuel now accumulating at reactor sites worldwide, but will ease the present conflict between the United States and Europe over international nuclear energy policy. America views reprocessing as a problem, a chief ingredient in the spread of nuclear weapons. Western Europe, on the other hand, sees it as a commercial venture and the vehicle for energy independence through breeder development.

Europe and America can probably agree that international storage meets both their objectives. It is superior to unsupervised storage at reactor sites. If the fuel can eventually be reprocessed, the Europeans will be satisfied; and if it delays the actual occurrence of reprocessing, the Americans will be satisfied. The United States might hope that it can convince Europe to forego the breeder, and thus reprocessing, if given "breathing time" by supervised international storage. And Europe, which will not have to start reprocessing until four or five years before scheduled breeder start-up, may see interim international storage as a good way to deflect American pressure.

While spent fuel storage is, in part, a temporizing measure, the time gained offers opportunities for addressing one of nuclear energy's most difficult quandaries – waste disposal. No one has yet devised a universally acceptable technique for final disposal of radioactive wastes. Although each country argues the merits of the method that best suits the geologic formations available in its territory, inadequate disposal anyplace could affect many countries. Countries with modest nuclear programs are looking to their supplier countries or to international organizations to dispose of their waste, and some are advocating a halt to nuclear construction until disposal methods are sorted out. Internationally supervised spent fuel storage extending over decades would buy time to design and implement international waste disposal arrangements.

International spent fuel storage would also aid the nuclear power industries of developed countries. There is a growing trend in nuclear countries toward requiring some plan for ultimate fuel disposition as a legal prerequisite for reactor licensing. If properly implemented, an international spent fuel storage regime could provide that avenue for final disposal.

Interim storage of spent fuel under international supervision would relieve the reactor licensing impasse and buy time for developing an international consensus on waste disposal and reprocessing. The fuel depositories could be structured to discourage developing countries from acquiring their own reprocessing facilities. They would promote economies of scale in reprocessing and permanent waste disposal. In short, spent fuel storage would both deter proliferation and aid the peaceful, profitable use of nuclear energy for electric power generation.

POSSIBLE APPROACHES TO SPENT FUEL STORAGE

There is a whole spectrum of systems for international spent fuel storage. At the lenient end is the starting or improving of multilateral supervision of spent fuel at the reactor sites. Although this would require the least change in international nuclear energy institutions and would alarm very few governments, it is the least effective method for containing proliferation. The acceptability and effectiveness of such arrangements must, however, be examined carefully in light of recent and ongoing advances in remote safeguard surveillance techniques. Such multilateral supervision might not be optimal for larger amounts of fuel over the longer term; but it might provide the best of all possible systems where interim management is necessary, or where small-scale diversion is possible.

At the other end of the spectrum of national acceptability is the return of spent fuel to the nuclear weapons states, the nuclear supplier states, and other OECD member countries. This policy has already been implemented by the Soviet Union, which requires the return of all spent fuel from countries purchasing its reactors; and the United States recently announced that it would accept foreign spent fuel, on a limited basis, in exchange for a one-time storage and handling fee. Such return arrangements can be extremely effective nonproliferation tools, and should be instituted in all possible situations; but they may be the least acceptable to both host and participant nuclear nations. Public opinion in potential host states is generally strongly opposed to accepting foreign spent fuel or waste, and public opinion and governments in many potential participant states are reluctant to turn spent fuel over to other states. The highly industrialized countries are especially problematic hosts. They not only give no credit for unrecovered fuel value in the plutonium and uranium, but they charge for storage and disposal.

Less-developed countries (LDCs) with existing nuclear power programs would be reluctant to send their spent fuel to a facility located in a country with huge stockpiles of nuclear weapons. Although some LDCs might accept a reasonably inexpensive regional international storage system, few, if any, would turn their spent fuel over to the developed countries. Therefore, facilities located in non-nuclear weapons states are a possibility. Arranging spent fuel return for countries where nuclear power programs are now in the planning stage stands a much greater chance of success. A very important model for such countries is the direct spent fuel return arrangement between the United States and Egypt, which is part of a broader nuclear cooperation program currently awaiting approval by the U.S. Congress. The spent fuel return and reprocessing arrangement between the Soviet Union and all countries purchasing its reactors provides another potential model.

The political complexities of international spent fuel management mandates a mixture of bilateral and multilateral arrangements. One should not rule out a broad international storage arrangement under the auspices of the International Atomic Energy Agency (IAEA). Although IAEA standards should be developed and applied to all fuel management schemes, separate regional and bilateral arrangements are worthwhile and should be encouraged.

Several forms of international spent fuel storage arrangements fall on the spectrum between multilateral supervision of national facilities and the return of fuel to developed countries. Addressing the single but vital factor of site location, there are four possibilities; these are listed in order of increasing nonproliferation effectiveness: LDCs without nuclear industries or nuclear power programs would be one option; LDCs with nuclear power programs would be another; a third might be an island, placed under some degree of international control with close IAEA supervision; and a fourth option would be to locate the sites in nuclear weapons states, nuclear supplier states, or OECD countries.

Options one and two might not be suited to any facility that includes reprocessing. Most LDCs could not provide sufficient local expertise for all back-end fuel cycle operations, and nonproliferation effectiveness would be low. A strict storage arrangement might be significantly different; several locations under options one and two might possess adequate technological capacity and nonproliferation effectiveness. Furthermore, the prestige of hosting an international nuclear facility might produce host countries in options one and two.

Options one and two offer the important advantage of possibly providing a politically and legally acceptable local waste disposal site. While co-location of disposal and storage sites – along the lines of the storage/reprocessing/disposal site intended at Gorleben, West Germany – may not be possible, or even desirable, longer-term storage with some possibility of local disposal is much more likely in LDCs than in developed countries.

Option three, an internationally controlled territory or island, would probably not offer a geologically acceptable site for local waste disposal. Medium-term storage without specific arrangements for final disposition might work, however, making this option a distinct possibility.

The developed nuclear countries of option four offer some of the best geological possibilities for local waste disposal, but political opposition would prevent the disposal of large amounts of foreign radioactive waste. The negative public attitudes towards importing foreign spent fuel for disposal has been highlighted recently by the detailed waste management contracts for reprocessing drawn up by England and France. Among other provisions, the contracts allow those countries to return waste to the customer nations. The Soviet Union is apparently adopting a similar policy; and other nations planning commercial reprocessing operations, such as West Germany and Belgium, have indicated their intentions to follow the same path. The return of reprocessing waste to customer nations makes no sense for security, health, or safety. Reprocessing countries readily admit this; but their insistence upon fuel return demonstrates the depth of public opposition to local disposal of foreign wastes, and the governmental-industrial sensitivity to further potential disruptions of national nuclear programs.

A PROGRESSIVE SPENT FUEL STORAGE REGIME

Outlines of the System

It would be inappropriate in a chapter that stresses the virtue of gradualism to advocate an immediate transition to a system of large-scale, centralized spent fuel repositories under international control. Progress must come in small, simple, careful steps. Indeed, if one of the intermediate steps adequately serves the international community, or if new forms of energy permit, the later stages might be abandoned.

Thus a realistic spent fuel storage regime must grow in carefully planned stages, with activity on various fronts occurring simultaneously. Progress on one front cannot be independent of the others; each step and each stage of the regime as a whole must aid nonproliferation and be able to mesh with the existing nuclear energy system. International organizations change slowly, if at all, and work on the longer-range elements of the spent fuel management program should begin immediately.

The least complex segment of a spent fuel management system would be reinforcing spent fuel return policies. The USSR's spent fuel return policy (see Nathanson) should be continued and supported by the West. Concerned nations should lobby with the U.S. to implement and expand its policy of accepting foreign spent fuel. This policy is currently limited to those countries that pose a real proliferation risk; it should be extended to states with small nuclear programs that want to avoid reprocessing and waste disposal. Finally, new vendors of fuel should be encouraged to require return, through leasing or some other mechanism, rather than transferring title and control outright.

The next, and more complex, stage of the spent fuel system would be to stiffen institutional supervision of at-reactor spent fuel storage. Maximum use should be made of existing at-reactor capacity by reracking and densification. Pools for reactors under construction should be redesigned to add capacity, and international agencies should increase their supervision. Specifically, all stored spent fuel should be subjected to a regime with real-time, or near-real-time, safeguards (see Miller). Close international monitoring is essential.

A further stage would be to put spent fuel in unused storage space at commercial reprocessing facilities. Fuel would be held under "storage only" contracts, renewable or convertible into reprocessing contracts as circumstances develop. Here, too, expanded international supervision would be vital.

The final and most complex stage of internationalized spent fuel storage would require building special storage facilities. These centers would offer a spectrum of services, including storage with return on demand, storage for a set period that would be later renewable or convertible into permanent storage, and permanent storage with no right of return. Such facilities would most likely have sufficient economies of scale to overcome transportation deficits, and would thus cost less than constructing further national storage sites (see Greer and Dalzell).

These segments are separable but complementary parts of an entire spent fuel management system. Not all are needed, or possible, at once. However, work should begin immediately on all to assure their availability when needed.

Relationship to Reprocessing

Any spent fuel returned to a nuclear fuel exporter under a return or lease policy, or shipped to the U.S. pursuant to its takeback offer, is then subject to those countries' reprocessing policies. Nevertheless, a spent fuel management system need not concern itself with reprocessing. If storage were strictly supervised, no restrictions on eventual reprocessing would be necessary, especially for the existing plants of supplier countries or future multinational plants. One would expect, however, that the fuel supply contract would require depositors to renounce construction of new national facilities. The spent fuel sent specifically for storage at reprocessing facilities could be reprocessed as storage contracts expire and breeder fuel requirements develop. To minimize transport, any eventual reprocessing could occur in the same facility where the fuel is stored.

Reprocessing would become problematic only in newly-constructed storage facilities because the location and oversight system selected for a new storage facility could be either conducive or hostile to future reprocessing. A spent fuel management regime should try, for reasons of political acceptability, to remain neutral toward reprocessing; whatever organizational scheme and site are selected, they should neither preclude nor require the later addition of reprocessing.

Institutional Development

Expanding spent fuel return policies would not require complex changes in international institutions. Encouragement and exhortation in bilateral and multilateral negotiations, such as the London Suppliers Group, and – if returns were to be internationally monitored – assistance from the IAEA safeguards system would probably suffice. The technical assistance required for expanded at-reactor storage, preferably through the IAEA, would present no institutional barriers beyond expansion of budgets. Providing storage space at existing reprocessing plants would mandate forging some new bilateral relationships.

Implementing near-real-time safeguards, however, would require amending existing safeguards agreements significantly and changing the administration of the IAEA safeguards system. The IAEA safeguards division would be required to handle unfamiliar technology and new functions. Nevertheless, none of these reforms would necessitate the founding of new international organizations.

Constructing new centralized facilities devoted exclusively to storing spent fuel would, however, demand new institutions, their complexity varying with the needs of each region. Details of the new arrangements cannot be foreseen with any certainty, but some general characteristics can be set forth.

NEW INSTITUTIONAL ARRANGEMENTS FOR CENTRALIZED STORAGE

Organizational Functions

An organization created to manage spent fuel repositories should exert control at three levels. First, it should exercise operational control over the functions of the repository; this would include not only the day-to-day management, but also the financial management, the design, and the construction of the facility. The second level of control should be regulatory. This function would include both safeguards installation and management; it would specify and execute radiological safety standards in the design, construction, and operation of the facility. Third, the new organization should have some independent political authority to adjust the competing interests of members, to regulate the admission of new participants, to establish relations with nonparticipant states and international organizations, and to deal with host country matters. The primary question is how to distribute such functions among constituent parts of the institutional system.

Existing International Organizations

Existing international institutions should be enlisted to the maximum extent possible to manage new spent depositories. If these institutions are imperfect, the proper remedial course is to strengthen them; the creation of competing organizations would further fragment the international order.

The limitations of existing international structures, however, must be recognized. International institutions designed to cope with international technological problems are considerably less structured and weaker than national organizations directed to the same purpose. International institutions can exert considerably less power over constituents and are much more susceptible to political wrangling than their national counterparts. International organizations can be classified according to purpose, type of instrumentality, and function performed; by determining degrees of complexity along these three dimensions, they can be categorized in a three dimensional "policy space."(5) Existing international organizations cluster at the "low" ends of all three scales; there are very few examples near any extreme and none near the extremes of all three dimensions. The IAEA is probably the most advanced, and a centralized spent fuel management regime would push the limits even more.

Provision for Progressivity

Previous designs for institutions to handle the international fuel cycle have envisioned a strong, autonomous organization empowered to store, reprocess, and dispose of spent fuel.(6) It is clear, however, that a much less-developed organization would suffice for an initial venture into spent fuel storage.(7) Creating a spent fuel organization simply to store spent fuel – without the powers and duties of reprocessing and reallocating fuel value and disposing of waste – would simplify the legal and organizational barriers enormously. Any additions to the initial structure could be developed as needs and experience dictated.

Some argue that the repeated restructuring of the spent fuel regime is too onerous to be practical, but these difficulties would be more than offset by the value of what experience teaches. INTELSAT demonstrates that a successful international technological venture can start with a temporary structure, and that it is possible, and probably even necessary, to redesign such an organ to meet unforeseen challenges and profit from experience.(8) If experience with international technological institutions over the last twenty years divulges a general lesson, it is that initial structure cannot and does not anticipate the actual problems that arise in the course of time.(9) Incrementalism would not only prompt a structure requiring the least amount of time and effort to establish, but would avoid serious errors farther along.

Geographic Structure

Any spent fuel repository would have to conform with international standards for safety and nonproliferation, but no single facility would be likely to attract universal membership. Regional or multinational repositories are more feasible. Other areas of international cooperation demonstrate that regional approaches frequently offer significant advantages. Even the specialized operating agencies of the UN work through regional offices or individual states.(10)

Economic arguments also militate for regional fuel storage centers. Economies of scale are strong for capital costs and significant for operating expenses. Not only are there high lease charges for spent fuel casks, but fuel transport itself is costly; these costs increase with transport distance and would be especially high if all fuel were to be returned to Western Europe. Economies of scale might equal or exceed the increased costs of added transportation charges for a regional arrangement.

Organizational Structure

Designing an organization with a workable distribution of powers is a major challenge. In Europe and many other countries, the normal framework is a public corporation. These take various forms: some are chartered under the law that governs private corporations and the government has partial or total stock ownership; others are created by special statute and are directly controlled by a government ministry, thereby dispensing with the fiction of a corporate intermediary.

The public corporation has been extended to the international sphere with varying degrees of success. The international public corporation can fall prey to the same government interference and dependence on annual appropriations which often compromise national public corporations. It does, however, provide for greater efficiency than other organizational structures and for private sector participation.

Experience with International Organizations

International spent fuel storage is still so abstract a prospect that analyzing how other international bodies have divided up power and responsibility would be fruitless. We can, however, glean some relevant general lessons from the experience of other international organizations.

The two organizations most similar to the proposed spent fuel depository are INTELSAT and Eurochemic. INTELSAT has encountered considerable political difficulty in its short lifetime because of failure to anticipate difficulties and accommodate conflicting political interests. On the one hand, U.S. interests wished to extend their domination of global communications through the technological superiority of Comsat, a quasi-public U.S. corporation. Other countries wanted access to the new technology to gain a foothold in international communications. The American monopoly on technology encouraged other countries to request – and get – a more democratic organization.(11) Eurochemic began as a commercial nuclear fuel reprocessor. Various participating countries burdened it with research and development functions and it eventually collapsed under its own weight.(12)

These two examples suggest that the hasty development of a complex organizational structure is apt to cause trouble; a more tentative approach works better. They also indicate that failure to recognize from the start that participants will have conflicting purposes may doom the endeavor.

Legal Problems

The precise role of the IAEA in managing an international spent fuel repository is unclear, in part because some countries dispute the agency's legal authority to require deposits of spent fuel. Article XII A.5 gives the Agency authority to require the deposit of excess "special fissionable materials." Although the Article XX definition of "special fissionable material" appears to include spent fuel containing plutonium, several major Agency members disagree with this interpretation, citing the negotiating history of the statute in support of their position. The IAEA does have clear authority, under Articles IX and XI, however, to join in voluntary arrangements. Thus, the Agency may still be able to play some significant role in a centralized repository.

The IAEA should also formulate the technical regulations for spent fuel storage – proliferation standards and safety standards for design, construction, and fuel transportation – and the ways to enforce them. IAEA regulations would win international approval quickly. Violators could be punished by withdrawing IAEA technical support, financing guarantees, and fuel assurances. IAEA spent fuel regulations could be made internally binding on participant states, and could, therefore, be enforced through domestic courts. This could be achieved either by a general convention signed by all members, or by the present system, which has individual states adopt the IAEA rules as part of domestic regulations.

Devising criteria for discharge of fuel on deposit in the facility will be complex. Some participants will want to be able to recall their fuel "at will"; some will wish to "bank" it for a set period; and some will seek to deposit spent fuel indefinitely or permanently. On the other hand, the repository may wish to reserve the right to return spent fuel if all other disposal avenues are blocked.

Headquarters and status-of-personnel agreements with host countries may be problematic. These problems could be avoided, of course, if the repository were affiliated with an existing regional political treaty organization such as Euratom or ASEAN.

With respect to legal liability, it is clearly unacceptable to limit the regime's liability to the value of its assets. Liability, especially for nuclear accidents, will have to be well insured against or passed along to member countries automatically.

SITING INTERNATIONAL SPENT FUEL STORAGE FACILITIES

Reaching agreement on a multinational spent fuel storage facility requires finding a country both willing to act as host and acceptable to potential participants, especially those with strong nonproliferation interest. These problems are closely enough related to justify using one set of site selection criteria. These criteria have been developed on the basis of extensive research, discussions in various seminars, and interviews with many officials, environmentalists, and industrialists in the United States, other countries, and international organizations. Although tightly enmeshed with technical, economic and legal aspects, political considerations are the focus of these criteria. Similarly, while the national and international politics of such siting intertwine, it is useful to analyze them separately.

National Political Criteria

1) One criterion is a country's attitudes toward nuclear nonproliferation, especially the degree of international access allowed to fuel cycle facilities. To be a serious contender for hosting a spent fuel storage and/or reprocessing facility, a country should clearly have a record of demonstrated support for nuclear nonproliferation. For example, Mexico ranks very high, while many would rank India and Pakistan quite low on horizontal proliferation attitudes; and the United States and the Soviet Union rank low on vertical proliferation attitudes. In addition, IAEA inspection of all facilities should be allowed, although this would not apply, at least initially, to military nuclear production plants in nuclear weapons states.

2) Internal political stability is also an important criterion. There is a direct correlation between the level of political stability in a candidate country and the degree of regional and international control required for an acceptable and effective arrangement. If the region is extremely stable, more internal disquiet is tolerable within the host country; thus, while the stability of an Australia or Costa Rica is not essential, some reasonable degree of internal cohesion is necessary. This requirement would disqualify many candidate site countries.

This criterion has three important components: (a) The focus should be on the nature of the country's government – the frequency and nature of changes in governments and the associated history of how reliably given governments have honored previous international commitments. The roles and powers of the party and interest group systems are particularly important, as is the timing of important elections over the next few years. A low ranking on this component of internal political stability may be offset quite readily by strong international commitments. (b) The effectiveness of the military services and internal police, especially in domestic security, and their political roles and powers should be ascertained. While participating countries could jointly employ some inspectors and monitor plant security, some minimum level of local military and/or police effectiveness is vital. Security requirements increase greatly for reprocessing and plutonium handling. Because all transportation links are important elements in the security equation, security effectiveness must be national, as well as local and regional. (c) The third component of internal political stability is the present and expected problems with organized criminal, guerrilla, and terrorist groups. The probability of terrorist attack has rapidly become a major factor in deciding where to locate nuclear facilities. It should be very carefully assessed for each potential spent fuel storage site, although high stability internally and regionally can partly offset problems with potential terrorism.

3) A crucial attribute for a potential spent fuel host country is the status of its energy and nuclear power programs. Vocal opposition from environmentalists is now a serious deterrent to new nuclear construction, and international trade of spent fuel may be particularly upsetting.

Countries with large nuclear power programs generally have the greatest need for assistance in spent fuel management; they also have the best organized opponents to nuclear energy. The governments and nuclear industries in these countries are particularly sensitive about antinuclear groups; they dread the possibility of creating new controversy and new opponents through local siting of international nuclear facilities. Even interim spent fuel storage facilities would be perceived as implying some responsibility, at least indirectly, for final disposal. How much independent authority a government has in nuclear matters must be examined. If a government is based on a fragile coalition, and its mandate to rule rests – as in Sweden, for example – partly on its nuclear energy views, very little enthusiasm can be expected for hosting international nuclear facilities. If a country's population is particularly sensitive to nuclear issues, as in Japan, Australia, New Zealand, or some of the Pacific Islands, the government's latitude may also be seriously restricted.

Thus, it may be easier, for reasons of domestic political acceptability, to site international spent fuel facilities in countries without commercial nuclear energy programs; countries with authoritarian regimes that will not tolerate environmental opposition to nuclear energy; or isolated areas, such as islands, turned over to international management. All three approaches deserve study.

4) Another consideration is whether effective economic incentives

can be offered to the host country's energy program. Special pricing, subsidies, fuel supply guarantees, and other incentives could be offered to the host country.(12) And, in itself, hosting the facility might be worth even more than money in the international prestige it would generate.

Potential host countries, especially LDCs, might be further enticed if their own nuclear power programs could receive transfusions of technology and services through the spent fuel facility. Taking the lead in nonproliferation might win some domestic political points for host countries with developed nuclear capabilities.

5) Geographic, geologic, hydrologic, climatic, and demographic suitability of candidate sites have to be considered. These technical criteria are much easier to satisfy for simple fuel storage than for reprocessing or disposal. An ideal site could support the type of centralized storage/reprocessing/disposal facility planned at Gorleben, West Germany, but such sites areextremely rare. Areas suitable for storage alone are more common; and, for the time being, they are desirable ends in themselves. West Germany is readying storage-only facilities at Ahaus, and the United States and other countries are poised to travel the same route.

Insisting that all storage sites meet the technical criteria for waste disposal could present two further problems. No country has yet found geologic formations thatseem acceptable for the final disposal of high-level radioactive waste or spent fuel, and demonstrations of operational repositories are at least a decade off. Permanent safe disposal may be impossible altogether in several countries with nuclear power and where possible, disposal of foreign wastes may be unacceptable. Furthermore, many of the potential fuel storage sites that are most attractive for political reasons – islands under international control or national territory convenient to transport – are unlikely to be suitable for final waste disposal.

International Political Criteria

1) International political acceptability is an important consideration. Deciding where to locate a spent fuel regime will depend in part on the social, cultural, religious, economic, and other affinities between potential host countries and participants. North-South and East-West relations will play a role here, as will countries' human rights records.

2) Closely related to the degree of internal political stability is the stability of the region and the potential host state's role in regional and international security systems. Regional rivalries and relative military strength must be assessed to assure the security of the spent fuel storage facility and its associated technical, industrial, and transport activities.

Rivals in two-country disputes – Pakistan and India, or Argentina and Brazil – will most likely veto locating any spent fuel facility in their foe's territory. However, rivalry may also produce benefits: Countries may vie for the prestige of hosting a sophisticated international facility. They may also decide that the benefits in eventual regional stability outweigh the cost of letting a foe be the host.

3) Another factor influencing a country's suitability as a host is its role in international nuclear markets. This is a complex market, as indicated by the following list of nuclear exports and exporters.

a) Uranium: United States, Soviet Union/Czechoslovakia, Canada, Australia, South Africa, France, Niger, and Gabon.

b) Uranium enrichment and fuel fabrication services: United States, Soviet Union, Eurodif (led by France), and Urenco (U.K., Netherlands, and West Germany).

c) Reactor vessels, steam supply systems and associated equipment: United States, Soviet Union, Canada, France, West Germany, Sweden, Japan, and Italy.

d) Spent fuel management services: United States and Soviet Union.

e) Reprocessing services, including interim spent fuel storage: France, United Kingdom, and West Germany (?).

f) Vitrification for high-level radioactive wastes from reprocessing operations: France and United Kingdom.

g) Final radioactive waste disposal services: United States (?), Canada (?), Soviet Union (?), Sweden (?), West Germany (?), and France (?).

h) Transport services for spent fuel, high-level wastes from reprocessing, or other radioactive materials: United Kingdom.

Countries exporting in categories (a) and (b) are at least entertaining the possibility of assuming responsibility for the later disposition of uranium or fuel they have exported; such responsibility could eventually be discharged, in part, through supporting, or even hosting, an international spent fuel storage facility. Countries able to offer contracts or guarantees for goods or services in categories (a), (b), (c), (g), or (h) might use these as incentives for countries to participate in spent fuel management arrangements. Countries selling services in categories (e) and (f) may compete with spent fuel management schemes to win additional long-term contracts for commercial reprocessing.

Two important caveats must, however, be attached to this argument that the atomic marketplace may be enlisted in the cause of spent fuel management. First, some countries in categories (a), (b), (c), (g), and (h), and especially the United States, are tightly constrained by law, politics, and economic philosophy against manipulating the marketplace.

Government-controlled nuclear services offer some flexibility in achieving foreign policy objectives, but even they must generally be offered to foreign customers on equal terms; most services come from private corporations which, though subject to governmental inducements, must make money on their activities.

The second caveat is that many governments that import material strongly dislike these imports being used as carrots and sticks for compliance on nuclear fuel cycle issues. The U.S. government, in particular, has to be extremely careful not to undercut its international proliferation position by appearing to be motivated by commercial objectives. Especially in the International Nuclear Fuel Cycle Evaluation, the U.S. must be above suspicion.

4) The fourth international political criterion for siting spent fuel storage facilities is the potential host country's associations with international organizations and nonproliferation treaties. This criterion is closely related to the national one on nonproliferation attitudes. It includes all memberships and relationships established with organizations such as IAEA, OECD/NEA, EURATOM, and other regional bodies involved in nuclear free zones, or energy regulation and development. Ratifications of the Nonproliferation Treaty, the Limited Test Ban Treaty, the Seabed Arms Control Treaty, and the various agreements on third party liability for nuclear damage are obviously important.

5) Finally, the geographic situation must be assessed. The leading element of this criterion is how close the country is to transportation routes to potential participant nations and to possible reprocessing and/or waste disposal sites. Geographic considerations become more important in direct proportion to the priority attached to security and health/safety risks from transportation. Careful attention must be directed to assessing the transport trade-offs between individual national storage and international storage, especially in the Pacific where distances are so vast.

NOTES

(1) Among the countries with nuclear power reactors in operation, under construction, ordered, or committed in a letter of intent, the following have not demonstrated any intention to have national reprocessing programs: Austria, Czechoslovakia, Egypt, Finland, East Germany, Hungary, Iran, South Korea (under strong pressure from the United States), Luxembourg, Mexico, Netherlands, Philippines, Poland, Rumania, South Africa, Sweden, Switzerland, Taiwan, and Yugoslavia. With the exception of Yugoslavia, which is developing an agreement with the United States for controlling all spent fuel produced in its Westinghouse reactor (U.S. approval of any reprocessing will apparently be required), the Eastern European countries are bound by bilateral arrangements to return spent fuel to the Soviet Union. Under a pending agreement, Egypt would return all spent fuel to the United States. Nations such as the Netherlands and Switzerland are investigating future reprocessing in France. Austria, Finland, Italy, South Korea,

Luxembourg, Mexico, the Netherlands, the Philippines, Spain, Sweden, and Taiwan have not expressed clear plans for the disposition of their spent fuel. At the least, Austria, Denmark (planning a future program), and Sweden have formally accepted or discussed the American offer to manage foreign spent fuel.

(2) General Report of the Scientific and Technical Committee, North Atlantic Assembly, International Secretariat, U110, STC(77) 8, presented by Georges Mundeleer, General Rapporteur, paragraph 110, p. 16.

(3) Nuclear Energy Policy Study Group, Nuclear Power Issues and Choices (1977).

(4) Manfred Hagen, An Assessment of World Reprocessing Capability and the Approach of the Federal Republic of Germany, Summary Report of the Fuel Cycle Conference 76, Atomic Industrial Forum, p. 69 (June 1976).

(5) The first dimension is the purpose of the organization, and the three classification points along the axis are (1) the acquisition of new capabilities, including R & D, (2) making effective use of capabilities which already exist, e.g., civil aviation, and (3) coping with the consequences of the use of capabilities, e.g., safeguarding nuclear materials. The second dimension is the type of instrumentality, with four possibilities along the axis with an increasing degree of control of national behavior: (1) a common framework for national behavior, (2) a joint facility coordinating national behavior, (3) a common policy integrating national behavior, and (4) a common policy substituted for independent national behavior. The third axis of the classification scheme is function performed: (1) informational, (2) managerial, and (3) executive. An additional variant of the international organization concerns classification of the task it performs: (1) It facilitates by planning activities to be carried out by member governments; (2) it enables by planning plus making decisions but not operating; and (3) it carries out operational tasks. An example of the application of this classification is the IAEA safeguards program, which performs an operational task, whose purpose is to cope with the consequences of nuclear technology, whose instrumentality is a common policy concerning safeguards matters and displacing national policy, and whose function is informational.

(6) C. R. Smith and A. Chayes, "Institutional Arrangements for a Multinational Reprocessing Plant," in A. Chayes and W. Lewis, International Arrangements for Nuclear Fuel Reprocessing (Cambridge, Mass.: Ballinger Publishing Co., 1977).

(7) Ibid., p. 151.

(8) S. Levy, "INTELSAT: Technology, Politics, and the Transformation of a Regime," International Organization 29:655 (Summer 1975).

(9) P. Strohl, Appraisal of Legal and Administrative Problems Concerning the Setting Up and Operation of Joint Projects in the Field of Energy R & D, OECD/NEA, at n. 2 (Feb. 20, 1975); Annex B to IAEA Regional Nuclear Fuel Cycle Centre Study Institutional-Legal Framework Aspects (July 1976).

(10) C. Okidi, "Toward Regional Arrangements for Regulation of Marine Pollution: An Appraisal of Options," Ocean Development and International Law 4:1,12 (1977).

(11) S. Levy, op. cit., pp. 655, 661.

(12) See note 9.

II

Regional Considerations and Possibilities

2 Eastern Europe and the Soviet Union

Melvyn B. Nathanson

INTRODUCTION

One waste product that comes out of a nuclear power plant, and one of the nuclear wastes most difficult to dispose of safely, is technology useful for building atomic bombs. A non-nuclear-weapons state can acquire the knowledge and experience to produce nuclear weapons by developing a civilian nuclear power industry and a cadre of scientists and engineers familiar with nuclear technology. Non-nuclear-weapons states that now have – or that plan to have – a substantial nuclear industry, but have not yet ratified the Nonproliferation Treaty, include Argentina, Brazil, India, Israel, Pakistan, and South Africa. A goal of both American and Soviet foreign policy is to prevent these and other countries from manufacturing atomic bombs.

Two types of thermal nuclear reactors are in commercial operation. The first is the light-water reactor fueled by uranium slightly enriched to about three percent U-235. This was developed first in the United States and is now manufactured in the United States, France, West Germany, Japan, Sweden, and the Soviet Union. The second reactor is the heavy-water reactor fueled by natural, unenriched uranium with about 0.7 percent U-235; this was developed in Canada – the CANDU reactor – and in Czechoslovakia. Both reactor technologies use a once-through fuel cycle in which uranium goes from a fuel fabrication plant through the nuclear reactor to spent fuel storage. Plutonium appears only in the spent fuel, and there it is mixed with highly-radioactive fission waste products and large quantities of U-238, the abundant, naturally-occurring isotope of uranium. It is not necessary to reprocess the spent fuel and extract plutonium. However, plutonium is the critical component of atomic weapons, and so a country that possesses both spent fuel and a reprocessing plant can easily "go nuclear."

Consequently, a partial solution to the problem of preventing the proliferation of nuclear weapons is to remove the spent fuel from national control, and to place it under international supervision, either globally or regionally. This paper considers the Soviet and Eastern

European aspects of a nuclear spent fuel storage regime. This is particularly important because the USSR is a military and political superpower with ambitious plans for commercial nuclear power within its boundaries and throughout Eastern Europe, and with an expanding program of nuclear exports to the Third World. The contractual obligations that the Soviet Union puts on the nuclear reactor fuel it sells abroad, and on the spent fuel produced in nuclear reactors it has exported, will affect the implementation of any spent fuel storage regime. Moreover, the International Atomic Energy Authority (IAEA) will be involved in the organization and management of spent fuel storage depots. The USSR is an active member of the IAEA, and Soviet approval will be necessary for IAEA participation. Finally, a Soviet-dominated spent fuel storage regime is already in operation in Eastern Europe, and this Soviet-bloc storage system is regarded by some as a model for other parts of the world.

In the next section Soviet policy on nonproliferation and arms control is briefly reviewed. The following two sections describe the commercial nuclear power programs in the USSR and in Eastern Europe. The final section analyzes the possibilities for Soviet and Eastern European participation in regional and global international spent fuel storage regimes.

SOVIET NONPROLIFERATION POLICY

Opposition to the use of nuclear weapons has been part of Soviet foreign policy and propaganda since 1945. This opposition was particularly evident in the years just following World War II when Moscow was profoundly inferior to the West in nuclear weapons and delivery systems, but superior in conventional forces. About 1960, after Moscow had produced atomic and hydrogen bombs and had begun to fear the acquisition of nuclear weapons by such potential enemies as West Germany and China, the Soviet Union again accelerated its support of nuclear nonproliferation proposals.

The history of Soviet policy in nuclear arms control and nonproliferation is an interesting one. On December 27, 1945, the United States, the United Kingdom, and the Soviet Union issued a communique urging the establishment of a commission to make specific proposals:

a) For extending between all nations the exchange of basic scientific information for peaceful ends;
b) For control of atomic energy to the extent necessary to ensure its use only for peaceful purposes;
c) For the elimination from national armaments of atomic weapons and of all other major weapons adaptable to mass destruction;
d) for effective safeguards by way of inspection and other means to protect complying states against the hazards of violations and evasions.(1)

The United Nations General Assembly, in its first session on January 24, 1946, unanimously approved the establishment of this commission. On June 14, 1946, the American representative to the United Nations

Atomic Energy Commission, Bernard Baruch, proposed the creation of an International Atomic Development Authority that would have "managerial control or ownership of all atomic-energy activities potentially dangerous to world security," as well as the "power to control, inspect, and license all other atomic activities."(2) According to the Baruch plan, when this control system was in effective operation, "manufacture of atomic bombs shall stop" and "existing bombs shall be disposed of."(3) This plan, which was proposed at a time when only the United States possessed atomic bombs and which appeared to give the United States a perpetual monopoly on the technology and experience to manufacture nuclear weapons, was unacceptable to the USSR. On June 14, 1946, the Soviet representative to the United Nations Atomic Energy Commission, Andrei Gromyko, presented an alternate plan, which required a prohibition on the production and employment of nuclear weapons and the destruction of all existing weapons, to be followed by the establishment of methods to ensure that atomic energy would be used only for peaceful purposes.(4)

Neither the Baruch plan nor the Gromyko plan was adopted. The Soviet Union issued a succession of proposals for the international control of atomic energy and for reductions in conventional forces; these were always tied to a prohibition on the use of atomic weapons, but were never accompanied by inspection procedures adequate to guarantee compliance. On September 24, 1948, the Soviet Union introduced a resolution in the General Assembly calling for "the reduction by one-third during one year of all present land, naval, and air forces" and "the prohibition of atomic weapons as weapons intended for aims of aggression and not for those of defence."(5) Another Soviet General Assembly resolution of December 12, 1950, called for

> a draft convention for the unconditional prohibition of the atomic weapon and a draft convention for the international control of atomic energy, bearing in mind that both these conventions should be concluded and brought into effect simultaneously. . .(6)

The General Assembly rejected both resolutions.

No matter how strongly Washington might have desired the international control of atomic energy, the United States could not accept an absolute prohibition on the use of nuclear weapons in war. Because the USSR maintained an enormous superiority in conventional forces, Washington believed that nuclear weapons would be necessary to repel a Soviet invasion of Western Europe. This Soviet superiority in conventional forces, together with Moscow's refusal to allow on-site inspection in the USSR, made Soviet proposals for the proportional reduction in the size of armed forces unacceptable to the United States. Thus, on November 12, 1948, Frederick Osborn, adviser to the United States delegation at the General Assembly, stated:

> The Soviet states apparently have available combat troops at least five times more numerous than those of all Western European states put together. . . A reduction of one-third would

> not change the disproportion in Soviet armies. . . How can we know which of the nations should reduce or have reduced their arms by one-third or by one-half or by three-fourths without basic knowledge on which to make our decision, and without real knowledge of what goes on behind the Iron Curtain. . . . The Soviet Union alone is working behind a veil of secrecy. How then can the rest of the world disarm?(7)

By 1955, the Soviet Union had acquired nuclear weapons and was maintaining large nuclear research programs. At the same time, the Russians' deep historical and hysterical fear of Germany and their growing disenchantment with Peking led the Soviet Union to adopt nonproliferation policies in order to prevent the acquisition of nuclear weapons by West Germany and China. Indeed, many political scientists believe that Soviet nonproliferation propaganda against Bonn was intended to justify Moscow's refusal to give nuclear weapons to Peking. Thus, Bloomfield, Clemens, and Griffiths write:

> Soviet anxieties about nuclear proliferation to Germany were probably matched or surpassed by a desire to keep nuclear weapons from China, and the entire burden of Soviet propaganda against nuclear proliferation in the West could also be used to justify the Kremlin's denial of military assistance to Communist countries, above all to China, especially once it had been decided to deny China effective nuclear assistance.(8)

A frequent Soviet nonproliferation proposal was the prohibition of all nuclear-weapons tests. On January 14, 1957, a Soviet General Assembly resolution called upon "states conducting atomic and hydrogen weapons tests to discontinue them forthwith."(9) The Soviet Memorandum on Partial Measures in the Field of Disarmament, transmitted to the General Assembly on September 20, 1957, stated:

> The Soviet Government continues to insist on the necessity of reaching an agreement on the discontinuance of tests of atomic and hydrogen weapons, without making agreement on this subject conditional on the reaching of agreement on other aspects of disarmament. . . A positive solution of this problem would lead to a considerable improvement of the international situation and would greatly enhance the prospects of reaching agreement on other aspects of disarmament and of putting an end to the current armaments race.(10)

However, the Soviet Union still opposed verification procedures to insure compliance with its disarmament proposals:

> The Soviet Government maintains its view that aerial photography can neither prevent surprise attacks nor ensure the necessary control over disarmament. . . No peace-loving State can agree to the aerial photography of the whole of its territory without jeopardizing its security.(11)

The Soviet Union supported the creation of a geographically-defined nuclear free zone proposed by Polish Foreign Minister Adam Rapacki to the General Assembly on October 2, 1957:

> . . . the Government of the People's Republic of Poland declares that if the two German States should consent to enforce the prohibition of the production and stockpiling of nuclear weapons in their respective territories, the People's Republic of Poland is prepared simultaneously to institute the same prohibition in its territory.(12)

Also, the Soviet Union voted in favor of an Irish Draft Resolution on nonproliferation which was introduced in the General Assembly on October 17, 1958:

> The General Assembly . . . recognizing . . . that the danger now exists that an increase in the number of states possessing nuclear weapons may occur aggravating international tension and the difficulty of maintaining world peace . . . decides to establish an ad hoc Committee to study the dangers inherent in the further dissemination of nuclear weapons. . .(13)

The United States abstained from voting on this resolution, which contained no provision for verification.

On March 31, 1958, the Supreme Soviet of the USSR decreed a unilateral moratorium on the testing of nuclear weapons.(14) In a letter to President Eisenhower on April 4, 1958, Soviet Premier Khrushchev was explicit about the nonproliferation aspect of this decision:

> Today only three powers so far – the USSR, the U.S., and great Britain – possess nuclear weapons, and therefore an agreement on the discontinuance of nuclear weapons tests is comparatively easy to reach. However, if the tests are not now discontinued, then after some time other countries may become possessors of nuclear weapons and under such conditions it will of course be a more complicated matter to reach an agreement on the discontinuance of the tests.(15)

However, the Soviet Union still refused to allow international controls, and, therefore, Eisenhower declined to prohibit the further testing of American nuclear weapons. Khrushchev again urged him to join the test ban; "It is never too late for good deeds," he wrote.(16) On September 30, 1958, the Soviet Union resumed nuclear tests in the atmosphere.(17)

China is the great failure of Soviet nonproliferation policy. Soviet nuclear assistance to China began in 1955 and terminated in August 1960, when Moscow withdrew its technicians from China. Without Soviet help, China would not have been able to explode its first nuclear device in October 1964. Pollack has observed that "given the enormous surge of interest in the worldwide diffusion of nuclear technology and the possibility that this could lead to widespread proliferation of nuclear weapons capabilities, the lack of attention to the Sino-Soviet case is quite striking."(18)

The Chinese decision to acquire nuclear weapons seems to date from the mid-1950s.(19) Earlier, Mao Tse-tung did not consider nuclear weapons to be decisive military instruments: "The atomic bomb is a paper tiger which the U.S. reactionaries use to scare people," he declared. "It looks terrible, but in fact it isn't."(20) Mao soon changed his mind on nuclear matters; perhaps he decided that nuclear weapons were required to support Chinese claims and aspirations for major power status in world politics. Also, American threats to use nuclear weapons to end the Korean War in 1953, and to defend the Quemoy Islands in 1958, may have destroyed Mao's belief that the United States would never drop atomic bombs on China. Finally, production of nuclear weapons by Peking would strengthen Mao's doctrine of self-reliance in foreign and military policy.(21)

China opposed Soviet interests in detente, arms control, and, of course, nonproliferation. In 1955, the Soviet Union decided to share its nuclear expertise with "friendly" Communist countries, and promised China a research reactor and a cyclotron. By 1958, these facilities were in operation, and Chinese scientists were engaged in research at the Joint Institute for Nuclear Research at Dubna in the Ukraine. China also acquired a gaseous diffusion plant in Lanchow to produce enriched uranium. It is not clear exactly how much technical assistance the Soviet Union provided for the construction of this plant; but, because of the highly-sophisticated technology required for an enrichment facility, and because the physical characteristics of the gaseous diffusion plant at Lanchow duplicate those of a previously-built Soviet plant, it is likely that there was substantial Soviet aids.(22)

No one knows how much help Moscow promised Peking for the Chinese nuclear weapons program. China claims that it was promised a "sample" atomic bomb by Moscow:

> As far back as June 20, 1959, when there was not yet the slightest sign of a treaty on stopping nuclear tests, the Soviet Government unilaterally tore up the agreement on new technology for national defense concluded between China and the Soviet Union on October 15, 1957, and refused to provide China with a sample of an atomic bomb and technical data concerning its manufacture.(23)

Moscow has never confirmed or denied this assertion.

China had supported the Soviet demand for a complete prohibition of nuclear weapons: "Socialist countries do not want nuclear weapons. . . Nuclear weapons cannot be eaten. . . We maintain that a complete ban on nuclear weapons is an attainable goal. . ."(24) When Moscow relinquished its stand for total nuclear disarmament in favor of various nonproliferation objectives, Peking protested loudly and refused to comply:

> On August 25, 1962, . . . U.S. Secretary of State Rusk had

> proposed an agreement stipulating that, first the nuclear powers should undertake to refrain from transferring nuclear weapons and technical information concerning their manufacture to non-nuclear countries, and that, second, the countries not in possession of nuclear weapons should undertake to refrain from manufacturing them. . . . The Soviet Government gave an affirmative reply to this proposal of Rusk's. . . . (The) Chinese Government hoped the Soviet Government would not infringe on China's sovereign rights and act for China in assuming an obligation to refrain from manufacturing nuclear weapons.(25)

On August 5, 1963, the Partial Test Ban Treaty was signed in Moscow by the United States, the United Kingdom, and the Soviet Union; nuclear explosions were outlawed in the atmosphere and under the sea, but were permitted underground. Peking asserted that this tripartite treaty favored the United States:

> By omitting the prohibition on underground nuclear tests, the tripartite treaty legalizes such tests and makes it easier for the United States to improve its strategic nuclear weapons, develop tactical nuclear weapons, conduct nuclear blackmail, and prepare for "limited nuclear wars."
>
> . . . Since the United States has more experience in underground nuclear tests, to ban other forms of testing while preserving underground tests will retard Soviet progress and enhance the superiority of the United States.(26)

Peking charged that Moscow's purpose was not to keep atomic weapons out of West Germany, but to keep China and other socialist countries from possessing these weapons.

> Formerly we thought the Soviet leaders were genuinely afraid of the West German militarists coming into possession of nuclear weapons. Now we see that they trust U.S. imperialism and think it does not matter if the West German militarists possess nuclear weapons provided they are under the control of the United States. . . The real aim of the Soviet leaders is to compromise with the United States in order to seek momentary ease and to maintain a monopoly of nuclear weapons and lord it over the socialist camp.(27)

The Soviet Government rejected China's accusations:

> It would be naive, to say the least, to assume that it is possible to conduct one policy in the West and another in the East, to fight with one hand against the arming of West Germany with nuclear weapons, against the spreading of nuclear weapons in the world, and to supply these weapons to China with the other hand. . . The position of the Chinese government, set forth in the

> statement of August 15, can be understood only as meaning that the Chinese leaders do not care how nuclear weapons spread among the capitalist countries so long as the leaders of the People's Republic of China get a chance to lay their hands on a nuclear bomb and see what it is like.(28)

China replied:

> (The) Soviet leaders have colluded with the U.S. imperialists in an effort to force China to undertake not to manufacture nuclear weapons. . .(29)
>
> In the eyes of the Soviet leaders, the whole world and the destiny of all mankind revolve around nuclear weapons. Therefore they hold on tightly to their nuclear weapons, afraid that someone might take them away or come to possess them, and so break up their monopoly. They are very nervous.(30)

After China exploded its first nuclear device in 1964, West Germany became the critical country in Soviet nonproliferation policy. The exclusion of nuclear weapons from the control of Bonn is a major goal of Soviet foreign policy.(31) Moscow could not prevent China from manufacturing nuclear weapons; it does not want its Eastern European allies to possess them; it would have preferred that India not have tested a "peaceful" nuclear explosive device; it would protest vehemently against a nuclear explosion by Israel, South Africa, or Brazil; but atomic bombs under West Germany control are absolutely inadmissible. Soviet Premier Kosygin expressed this clearly in a London press conference in 1967:

> As for the FRG, I must say that it will have to join the agreement on nonproliferation whether it wants to or not. We will not allow the FRG to have nuclear weapons and will take all measures to prevent it from obtaining the possibility of possessing these weapons. We say this with full determination.(32)

As long as Moscow could hope to curb China's nuclear ambitions, it was restrained in its nonproliferation proposals. After the withdrawal of Soviet technicians from China in 1960, but before China's first nuclear test, came the first substantial agreement between the United States and the Soviet Union on nonproliferation, the Partial Nuclear Test Ban Treaty. Neither superpower would forego the further refinement of its own nuclear arsenal by outlawing underground tests. By eliminating above-ground nuclear explosions, the testing of a nuclear device by a non-nuclear signatory to the treaty was made more difficult. But not impossible; India ratified the Partial Nuclear Test Ban Treaty and exploded its first nuclear device in 1973.

The Treaty on the Nonproliferation of Nuclear Weapons (NPT) was signed by the United States and the Soviet Union on July 1, 1968. The United States, the Soviet Union, the United Kingdom, France, and China were the five countries that then possessed nuclear weapons. The Nonproliferation Treaty sought to prevent additional countries from acquiring such weapons.(33) Reversing their usual position on outside inspections, the Russians insisted that IAEA safeguards be imposed on all nations signing the Nonproliferation Treaty. The Soviet Union put pressure on countries to ratify the treaty. All of the Eastern European countries – with the exception of Albania – signed and ratified the Nonproliferation Treaty. Under Soviet pressure, almost all the Arab nations signed, even though Israel refused to sign.(34) China, of course, did not sign.

The NPT provided the Soviet Union with a convenient excuse to refuse requests for nuclear weapons, even from countries such as Cuba and Egypt that had not ratified the treaty. Indeed, Moscow has insisted that Havana and Cairo will have to accept IAEA inspections and safeguards in order to acquire Soviet-made nuclear power reactors.(35) However, the Soviet Union has not criticized all of the near-nuclear states that have refused to ratify the NPT; it has criticized only those that are opposed to Soviet foreign policy.(36) In particular, Moscow has not condemned the Indian nuclear explosion.

Soviet nonproliferation policy is opportunistic. As a conservative superpower, Moscow would prefer to maintain its nuclear duopoly with Washington; thus it generally opposes the spread of atomic weapons. Yet this opposition is not absolute; in particular cases, other Soviet foreign policy goals can – and do – take precedence over nonproliferation.

NUCLEAR POWER IN THE USSR

The Soviet Union has large and rapidly-developing nuclear power programs. Although the world's first nuclear reactor – an experimental 5 megawatt (MWe) light water reactor – began operation in Obninsk near Moscow in 1954, the USSR built up its commercial atomic power industry slowly. In the tenth five-year plan (1976-80), one-fifth of all installed electrical generating capacity will be fueled by nuclear reactors, and ten new nuclear power stations are under construction.(37)

Moscow's increasing dependence on nuclear energy has a simple geographical explanation: Although the Soviet Union is extraordinarily well-endowed with all of the major fuels,(38) these resources are located in Siberia and in Asia, far from European Russia where 80 percent of the Soviet population lives. A.M. Petros' yants, Chairman of the USSR State Committee on the Utilization of Atomic Energy, has clearly expressed this predicament:

> The Soviet Union is one of the fortunate nations from the point of view of energy fuels. Thus, in the Soviet Union can be found, for example, about 45 percent of the world reserves of natural gas, more than 50 percent of the coal reserves, peat, and, especially, hydroenergy resources that are still far from fully-utilized.

> The Soviet Union is the world's only major highly-developed industrialized state that bases its economic development on its own resources of energy fuels.
>
> However, the natural energy resources of the USSR are distributed unevenly and far from the basic economic region of the country. The European part of the USSR, where a large part of the population lives, senses more and more a shortage of fossil fuels.(39)

The expense of transporting energy west across the Urals is enormous. Extraction costs of oil, gas, and coal from Siberian fields vary; but the social cost of building roads and housing, and bringing in food and other amenities is immense. Once the fields have been exploited, Siberian energy must be moved to its users through vast networks of oil and gas pipelines, railroads, and high-voltage electricity transmitters; the cost is great. The alternate possibility of moving the energy users east to the sources of supply is also extraordinarily costly because of the harsh natural conditions of Siberia: blizzards, permafrost, and swamps.(40)

Consequently, the current five-year plan calls for the construction of 67 to 70,000 MWe of new electrical generating capacity of all types; of this, 13 to 15,000 MWe or about 20 percent, would be nuclear power, as compared with 6.5 percent for the five-year period 1971-75. This will increase Soviet nuclear power to 19,400 MWe in the European USSR by 1980.(41) By the end of 1974, the USSR had only 3,700 MWe of installed nuclear capacity, compared to about 14,000 MWe in Western Europe and 30,400 MWe in the United States.(42) Construction of another dozen Soviet atomic power stations will be started in the 1980s.(43)

Until recently, Soviet nuclear engineers were trained at the Moscow Engineering-Physical Institute in Obninsk, a research center about 70 miles southwest of Moscow. Because of the growing nuclear power programs and an increasing need for trained personnel, a special nuclear engineering college is being set up in Obninsk. This college, the Atomic Power Institute, will enroll 3,500 students.(44)

The USSR is able to plan its atomic energy development without considering the repercussions from a popular ecology movement and without concern for environmental and safety niceties. The Russians do not regard the disposal of radioactive wastes as a serious problem.(45) There was, apparently, a large nuclear accident near Sverdlovsk in the Urals about 20 years ago, but it was never reported in the Soviet press and was mentioned only recently in the Western press.(46) Another explosion was reported at the Shevchenko fast breeder reactor in 1973.(47) Although ecological problems and environmental protection have recently entered both the Soviet consciousness and the tenth five-year plan, a recent review of environmental issues in the USSR did not mention nuclear energy.(48) At the 250th anniversary of the Soviet Academy of Sciences in 1975, the distinguished Soviet physicist Pyotr Kapitsa warned of possible dangers in the careless expansion of nuclear power plants.(49) Although the speech was later printed in the Bulletin of the Soviet Academy of Sciences, it was never mentioned in the press and remains one of the few cautionary statements made in the USSR on nuclear power.(50) More typical of Soviet attitudes is the statement of Mikhail Troyanov, Deputy Director of the Obninsk laboratory: "I don't see any difficulties in going to plutonium."(51)

Soviet commercial nuclear power stations utilize both thermal reactors and, increasingly, fast breeder reactors. Two types of thermal reactors are in service in the USSR: one is the pressurized light-water moderated and cooled reactor, the VVER reactor series; and the other is the uranium-graphite channel-type boiling water reactor, the RBMK reactor series. Both use uranium fuel slightly enriched in the isotope U-235.

Pressurized water reactors are the reactors most in use in the Soviet Union, and, indeed, in the world. There are four commercial nuclear power stations in the USSR that employ VVER reactors. At the Voronezh nuclear power station, there are four pressurized water reactors – one VVER-210, one VVER-365, and two VVER-440 reactors – that together produce 1455 MWe electrical power. In addition, a 1000 MWe reactor, the VVER-1000 is under construction. On the Kola peninsula near Murmansk, there is a power station with two VVER-440 reactors in operation. Before this station was built, hydroelectric resources supplied about 80 percent of the power to the Kola peninsula. Coal and oil were hauled across the tundra to supply the remaining power. The Soviet Central Committee, in a statement printed on the front page of Pravda, heralded the successful operation of the Kola nuclear power station, which has the world's first nuclear reactors located above the Arctic Circle.(52) The Armenian nuclear power station and the West Ukrainian station, near Rovno, each have one VVER-440 reactor in operation and another under construction.

The largest pressurized water reactor manufactured in the USSR, the VVER-1000 reactor, produces 1000 MWe. To increase the power beyond 1000 MWe would require much larger reactor bodies, and Soviet officials admit that they possess neither the technology to produce them nor the railway system to transport them.(53) Nor do the Russians have the sophisticated technology that would allow them to build the reactor bodies in pieces, transport the pieces to the reactor site, and assemble them there.(54) However, uranium-graphite reactors of high power do not require enormous reactor bodies. Uranium-graphite reactors also produce more plutonium than pressurized water reactors, and plutonium is needed not only for military purposes but also as the primary fuel for fast breeder reactors.(55) Thus, to provide a transportable large nuclear reactor and to insure sufficient fuel supplies for the Soviet commercial breeder program, Moscow is building uranium-graphite reactors able to generate 1000, 1500, and 2400 MWe of electrical power – the RBMK-1000, RBMK-1500, and RBMK-2400 reactors, respectively.

In fact, the two oldest commercial nuclear power stations in the USSR were equipped with uranium-graphite reactors. The first Soviet nuclear power station began operation in December 1958 with six 100 MWe RBMK reactors. This is a covert station that is used both to generate commercial power and to produce plutonium for weapons. The Russians identify it as "Siberian"; American intelligence sources place it at Troitsk, a town in the southern Urals.(56) The second Soviet nuclear power station is at Beloyarsk, near Sverdlovsk, in the Urals, and consists of two RBMK reactors of 100 and 200 MWe electrical power, respectively.

The Soviet Union is building a series of 4000 MWe nuclear power stations, each consisting of four RBMK-1000 reactors. These are located in Leningrad, in Kurchatov, near Kursk, in Smolensk, and in Chernobyl, near Kiev.(57) Another nuclear power station with two RBMK-1000 reactors is under construction in Kalinin. At Ignalinski in Lithuania, there is a 3000 MWe power station under construction with two RBMK-1500 reactors. A 2400 MWe RBMK reactor is being developed.(58) In addition, in the southern Ukraine, on the Bug River, the USSR is creating a new city with a power-engineering complex consisting of an atomic power station, a hydroelectric power station, and a pumped-storage station.(59)

A recent article in the Soviet Magazine Geografiya v Shkole (Geography in School) lists 13 more sites where nuclear power stations may be built in the 1980s.(60)

The Soviet Union is constructing not only enormous power stations in European Russia, but also tiny nuclear power stations in isolated areas far removed from conventional energy sources. In the arctic village of Bilibino, near Magadan and the Bering strait, a 48 MWe atomic power station consisting of four 12 MWe reactors provides electrical and thermal energy.(61)

Moscow believes that there are not sufficient resources of U-235 to support the development of nuclear power generated by thermal reactors requiring enriched uranium fuel. Consequently, the Soviet Union intends to mass produce fast breeder reactors whose fuel cycles can utilize the abundantly-available supplies of natural uranium and thorium.(62) The first Soviet breeder reactor, the BR-5, began operation at Obninsk in 1958. It was upgraded in 1972 to 10 MWe, the BR-10 reactor.(63) Since 1968, an experimental 50 MWe breeder, the BOR-60, has been operating in Dimitrovgrad.(64)

In 1973, the world's first commercial breeder reactor, the BN-350, began operation in Shevchenko, a remote desert oilfield town on the northeast coast of the Caspian Sea.(65) This is a loop-type, 350 MWe liquid metal fast breeder reactor that is designed to yield 150 MWe electrical energy, and use its remaining energy to convert 30 million gallons of salt water from the Caspian Sea into fresh water each day. In November 1973, there was an explosion at the plant while it was operating at 30 percent of capacity; it remains in use, although still not at full capacity.(66) Petros'yants claims that it now operates at 65 percent of capacity.(67) Convict labor was used to build Shevchenko.(68)

A 600 MWe pool-type breeder reactor, the BN-600, which is similar to the French Phenix and the British PFR, is under construction at Beloyarak.(69) It is expected that the pool-type breeder, which is simpler than the loop-type breeder, will become the standard Soviet breeder design.(70) Currently under development is the BN-1600, a 1600

MWe fast breeder reactor.(71) Kazachkovskii, et al. describe the Soviet program for breeder reactors.(72)

While the Russians have designed and built many different nuclear reactors, their standard reactor has been the VVER-440 pressurized water reactor. Produced at a plant at Izhorsk, south of Leningrad, this has been widely used not only in the USSR but throughout Eastern Europe.(73) Nevertheless, Moscow's ambitious nuclear power program requires an enormously-expanded production capacity for atomic reactors.

The top priority construction project of the tenth five-year plan (1976-80) and the 25th Soviet Party Congress is to build a new city, Volgadonsk, in southern Russia. Situated at a point where the Volga and Don Rivers come close together, Volgadonsk is scheduled to have a giant nuclear power plant factory, Atommash;(74) some parts of this complex are already in operation.(75) In the 1980s this factory is expected to produce 1000 MWe reactors at the rate of perhaps eight a year.(76) According to the Director-Designate of the Atornmash plant, Mikhail F. Taryelkin, the factory may eventually produce reactors of more than 1000 MWe capacity; improved railroad construction will enable the transport of large reactors.(77) Eventually, the plant may produce fast breeder reactors. However, construction of Atommash is now far behind schedule.(78)

The Atommash factory will have a massive impact on the Soviet energy economy. By the end of the century, nuclear power will produce most of Soviet electricity. Transport of Siberian oil and gas to European Russia will greatly diminish. East of the Urals, where the coal fields lie, coal-burning power plants will produce electricity. Oil and gas will be used for chemical feedstocks and for export.

Atommash will also increase the ability of the USSR to export nuclear reactors. The Soviet Union is certain that atomic energy will provide most of the world's future energy needs. V.S. Emelyanov, Deputy Chairman of the USSR State Committee on the Utilization of Atomic Energy, has declared:

> Thus, the world enters a new era in which basic energy will be derived from nuclear processes. This is unavoidable. There are, in fact, no other suitable sources of power able to meet the energy requirements of world society. . . The switch from organic fuel to nuclear fuel is well under way, all the basic scientific and technological problems having been solved.(79)

A.P. Aleksandrov, President of the Soviet Academy of Sciences, has made similar statements.(80) The Russians are now exporting reactors not only to Eastern Europe, but also to Finland, Cuba, and Libya. The Soviet Union and the Eastern European countries have formed two international economic organizations, Interatomenergo and Interatominstrument, to develop technology and produce components for nuclear reactors.(81)

Moscow is considering purchasing foreign equipment for use in Atommash, and has held preliminary talks with Japanese companies about building reactor components in Japan to Soviet specifications.(82) Negotiations are in progress for Mitsubishi Heavy Industries to export parts for the Soviet 1000 MWE reactor. On November 22, 1977, Japan and the USSR signed their first agreement for cooperation on the peaceful uses of atomic energy; priorities were set for research on fast breeder and fusion reactors, on current light water reactor technology, and on reactor safety and waste disposal.(83) The Soviet Union has also offered to provide uranium enrichment services to Japan, with payment either in hard currency or in Japanese-produced components for Soviet nuclear reactors.(84)

There are indications that the Soviet Union seeks to develop an export market for its reactors in the Third World.(85) The standard Soviet 440 MWe reactor may be particularly attractive in less developed countries. It is the smallest available light water reactor. The CANDU reactor produces about 650 MWe, and other Western and Japanese reactors produce about 1000 MWe; these are too large for the electrical power grids of most Third World nations. Thus, the USSR could hope to monopolize the market for nuclear reactors in the Third World.

When the sale of a 620 MWe Westinghouse reactor to the Philippines became uncertain because of Westinghouse's questionable financial arrangements with the Marcos regime, the wife of President Marcos claimed that the Soviet Union had offered to replace the Westinghouse plant with a complete Soviet plant, including reactor, and to assist the Philippines in developing its uranium resources.(86) If Mrs. Marcos' claim is true, it serves as an example of aggressive Soviet marketing of atomic reactors. If Mrs. Marcos' claim is false, it shows nonetheless that the Soviet nuclear export program is sufficiently aggressive to provide a credible threat to the American nuclear export industry; consequently, the claim could put pressure on the U.S. Congress to allow Westinghouse to finish construction of the Philippine atomic power station.

The Soviet nuclear power industry is booming while the American nuclear industry stagnates. Daniel Yergin has succinctly described the state of atomic power in the non-Soviet world:

> Nuclear energy? Projections for its contribution have been consistently lowered since 1973. A tangle of barriers stands in the way – cost, technical problems, environmental risks, doubts about safety, and, most recently, the dispute over nuclear proliferation. To put it simply, nuclear power is in the grip of a paralyzing stalemate.(87)

NUCLEAR POWER IN EASTERN EUROPE

The Council for Mutual Economic Assistance (CMEA) consists of the Soviet-dominated countries of Eastern Europe – Czechoslovakia, the German Democratic Republic (East Germany), Bulgaria, Poland, Hungary, and Romania – together with the USSR, Mongolia, Vietnam, and Cuba. This section describes the nuclear power plans of the East European members of CMEA, and also of Yugoslavia and Albania, East

European countries that do not belong to CMEA. Yugoslavia has been an associate member of CMEA since 1965; Albania ceased to participate in CMEA in December 1961, after the Sino-Soviet split.(88)

The East European members of CMEA participate in several energy-related enterprises with the USSR. The Druzhba (Friendship) petroleum pipeline goes from the USSR through Poland to East Germany, and a branch runs southwest to Czechoslovakia and Hungary. The pipeline has an annual capacity of 105 million tons. In 1976 the Soviet Union exported 70 million tons of oil to Eastern Europe, and plans to export the same amount through 1980.(89) In 1975, the Soviet Union sold oil in Eastern Europe at about one-fourth of the world market price.(90) Soviet oil prices rose to about one-half the world price in 1976,(91) and may soon reach world price levels. Also, in 1975, for the first time, Moscow exported more oil to the West for hard currency than to Eastern Europe.(92) The Soviet Union has clearly profited from high OPEC oil prices and from the energy crisis in the West.

By the end of 1978, the Bratstvo (Brotherhood) natural gas pipeline will be capable of supplying natural gas to all of the European members of CMEA. These countries are supplying the labor and materials to construct a branch of the gas pipeline from Orenburg to Uzhgorod.(93) European CMEA members also cooperate in the Mir (Peace) electric power grid, which connects Kiev in the western Ukraine with Eastern Europe. According to Radio Moscow,(94) about 80 billion kilowatt hours of electricity were exchanged during 1971-75 and about 25 billion kilowatt hours in 1977. Two CMEA organizations, Interatomenergo and Interatominstrument, work on nuclear reactor technology.

In the decade following World War II, the Soviet Union guarded its atomic engineering technology tightly. However, in 1955, it decided to share its nuclear knowledge with other Communist countries. The USSR Council of Ministers issued the following decree on January 18, 1955:(95)

> The Soviet Government, attaching great importance to the use of atomic energy for peaceful purposes, has decided to provide scientific-technical and industrial help to other countries for the establishment of a scientific experimental basis for the development of research in the field of nuclear physics and the use of atomic energy for peaceful purposes. The Soviet Government has offered comprehensive help to the People's Republic of China, the People's Republic of Poland, the People's Republic of Czechoslovakia, the People's Republic of Romania, and the German Democratic Republic. . . .
> The question was considered of widening the circle of countries to which the USSR could also provide assistance and help. . . .

The Soviet Union signed nuclear research assistance agreements with China, Czechoslovakia, East Germany, Poland, and Romania in April 1955, and with Bulgaria and Hungary in the summer of 1956. In 1955-56, the Soviet Union also signed agreements with Czechoslovakia, East Germany, and Hungary on the development of commercial nuclear power.(96)

It is not clear why Moscow decided to export nuclear technology to its East European and Chinese allies. In the early 1950s, the Soviet Union was a net importer of energy.(97) A world-wide power shortage was feared, and so the huge cost of a nuclear power program might have seemed preferable to an economic decline. Also, the Soviet Union might have wanted to harness the economic and intellectual resources of the East European countries for pilot research and development projects for the Soviet nuclear industry. In 1955, the USSR had no commercial power reactors in operation. Nevertheless, it agreed to provide Czechoslovakia with a 150 MWe natural-uranium-fueled heavy water moderated reactor, which was supposed to become operational in 1960. After long delays, at considerable cost to Prague, and without much Soviet technical assistance, this reactor finally went into operation in 1972.(98) Similarly, the first small East German power reactor, the AKW-1, was, apparently, used to test Soviet designs and equipment for the VVER-210 reactor under development at Voronezh.(99)

The Soviet Union never carried out the 1955-56 nuclear assistance agreements with Eastern Europe; the USSR simply did not possess the sophisticated nuclear power technology it had promised to deliver. The proliferation danger created by nuclear exports might also have frightened the Russians. Between 1955-58, Moscow provided China with an unsafeguarded 6.5 MWe reactor, and, perhaps, a gaseous-diffusion enrichment plant.(100) By August 1959, Soviet nuclear aid to China had ended, but the Chinese were still able to finish construction of the enrichment plant. Weapons materials obtained from this plant were used in the first Chinese nuclear explosion in 1964. The ease with which the Chinese were able to adapt civilian nuclear technology to military purposes must have disquieted Moscow, and further moderated its enthusiasm for transferring nuclear technology to Eastern Europe.(101)

A third reason why Moscow did not carry out its nuclear assistance pacts with Eastern Europe was the reversal in the Soviet energy situation; in the decade from 1955-65, the USSR acquired a growing energy surplus. According to official Soviet sources, gross exports of liquid hydrocarbons to Bulgaria, Czechoslovakia, East Germany, Hungary, and Poland rose from 2.2 million tons to 22.4 million tons.(102) Moreover, the price of Soviet oil was well above world prices at this time. The increasing supplies of liquid and solid fuels imported into Eastern Europe in the 1960s removed the urgency for rapid development of nuclear power. Then, in the 1970s, the energy situation reversed itself again. Soviet oil was sold to Eastern Europe below world prices; and there were predictions that the USSR would again become a net importer of energy, and that Eastern Europe would become an economic liability to the Soviet Union.

Fears that the USSR would soon be deficient in both fossil fuels and in the natural uranium to fuel its nuclear reactors has led Moscow not only to expand its own atomic power capacity, but also to allow its East European allies to plan construction of nuclear power stations utilizing both thermal and fast breeder reactors.(103) At an IAEA conference on nuclear power in 1977, a group of representatives of CMEA countries reported on expected nuclear power developments in Eastern Europe:

> The development of atomic electrical power engineering plays a growing role in the solution of the fuel-energy problem in CMEA. The Council for Mutual Economic Assistance has frequently emphasized the importance of an accelerated development of nuclear energy in the member-nations of CMEA, and the necessity of an increase in the effectiveness of cooperation in this domain. . . The most effective method to solve the fuel problem is the wide introduction of fast breeder reactors (FBR). . . . (By the year 2000) the share of FBR's in the structure of nuclear power engineering could reach 50 percent.(104)

What are the nuclear energy prospects for the nations of Eastern Europe?

Czechoslovakia

In addition to having the most sophisticated nuclear industry in Eastern Europe, Czechoslovakia has large uranium deposits. Since the end of World War II, the Soviet Union has had exclusive access to these deposits in Jachymov and elsewhere.(105) A highly-industrialized country with the world's third highest per capita energy consumption, Czechoslovakia has extensive hard coal reserves of about 3 billion metric tons with an additional 3 billion tons of brown coal and lignite.(106) Although this is enough coal for at least 100 years, some of the reserves are not accessible with present technology, and it is not economically feasible to mine much of the rest because of the sharp rise in the cost of hard coal production. The country has, therefore, become a net importer of solid fuels. Nearly all of Czechoslovakia's petroleum and natural gas are imported from the Soviet Union; in 1974, over 14.6 million tons of oil came from the USSR.(107) In 1955, when the Soviet Union decided to permit the development of nuclear technology in Eastern Europe, Czechoslovakia was the first CMEA country to sign an agreement with the USSR to build a nuclear power plant. This was for a gas-cooled, heavy water moderated reactor, the Bohunice lA, with 150 MWe electrical capacity. It was supposed to have been operational by 1960,(108) but finally went on line in December 1972, producing 112 MWe.

Czechoslovakia had hoped to become an independent nuclear-power state, producing its own heavy water, and mining its own uranium for natural-uranium-fueled reactors. The USSR might have been pleased, at first, that Czechoslovakia expended its national treasure to develop this technology, which the USSR did not then possess; but, eventually, Moscow discouraged thoughts of independence in Prague. The Bohunice lA is the only heavy water reactor in Eastern Europe, although a CANDU reactor may be built in Romania. Four Soviet 440 MWe pressurized water VVER-440 reactors are now under construction in Czechoslovakia; Prague also intends to build 16 more Soviet pressurized water reactors, each with 1000 MWe capacity.(109)

The German Democratic Republic

Until recently, the only domestic fuel in East Germany was lignite. In 1974, it provided 72 percent of domestic energy needs. Lignite was used as a fuel for power stations and as raw material for the GDR's growing petrochemical synthetic material industries. Petroleum and natural gas were imported from the USSR. However, there have been recent discoveries of natural gas, and the domestic output is increasing.(111)

The German Democratic Republic was the first CMEA country to operate a nuclear power reactor. The AKW-1, a 70 MWe pressurized water reactor, began operating in May 1966 in Rheinsburg.(112) The AKW-1, a Soviet reactor similar in design to the VVER-210, required enriched uranium fuel. The GDR has, perhaps, the richest uranium deposits in Europe, and may have intended at one time to build natural uranium reactors.(113) This intention was never carried out, and the GDR now has three VVER-440 reactors in operation – and a fourth under construction – in Lubmin.(114) Four additional VVER-440 reactors are planned. Construction has also begun on a new nuclear power station in the town of Stendal, west of Berlin, that will use the new Soviet VVER-1000 reactors to be produced at Atommash.(115)

One of East Germany's leading atomic scientists is Klaus Fuchs of the Dresden Atomic Research Institute. In an article in the magazine Energietechnik, Fuchs predicted that the GDR intended to buy fast breeder reactors from the USSR, possibly by the end of the 1980s.(116)

Bulgaria

There is a severe energy shortage in Bulgaria. A large part of its petroleum is imported, 10.5 million tons in 1974; most of this came from the Soviet Union, with a small amount from Egypt and Libya.(117) Bulgaria decided early to develop nuclear power, and, in 1969, began construction of a nuclear power plant in Kozlodoi, north of Sofia near the Danube. Two VVER-440 reactors now operate in Kozlodoi, and two more are under construction. In addition, Bulgaria plans to build four VVER-1000 reactors.(118)

Bulgaria is the CMEA country with the strongest political and economic bonds to the Soviet Union. Over 50 percent of Bulgaria's export goes to the USSR, and Bulgaria takes third place in the foreign trade of the Soviet Union.(119) In addition, thousands of Bulgarians are at work on construction projects inside the USSR. One Bulgarian journalist writes about the Bulgarian seventh five-year plan (1976-80):

> Integration between Bulgarian and Soviet mechanical engineering enterprises particularly expanded... (It) is planned to achieve an even greater expanded production of: machines and equipment for Bulgarian and Soviet atomic power stations.... The agreements on Bulgaria's participation in the construction of major projects on the territory of the USSR contribute to the

development of economic integration and the deepening of cooperation.(120)

Poland

Poland is a net exporter of energy. Its own extensive coal deposits provide 80 percent of its total energy consumption, and also yield considerable quantities for export. Natural gas production has increased, and uranium is mined in Lower Silesia.(121) Poland imported 10.5 million tons of oil in 1974, 97.5 percent of which came from the Soviet Union.

Poland had planned to develop a small, independent nuclear power industry using its own uranium to fuel natural uranium reactors. Poland also hoped to develop a nuclear ship propulsion system. Both plans were abandoned, in part because it was difficult to justify large nuclear research expenditures in an energy-rich country, and in part because of Soviet displeasure.(122) Poland now plans to build one Soviet VVER-440 reactor sometime in the 1980s.(123)

Hungary

Hungary is an energy-poor country. In 1974, it supplied about 65 percent of its energy needs; domestic productions accounts for only about 30 percent of its petroleum consumption.(124) Most of its needed oil is imported from the Soviet Union. According to Hungarian government officials, the "friendship price" of Soviet oil to Eastern bloc countries will equal the world price within the next two years.(125) Moreover, Moscow has ordered a freeze on increases of its oil exports to Eastern Europe. Hungary has recently opened the 10-million-ton-a-year Pan-Adriatic pipeline that will carry Middle Eastern oil from the Yugoslav coast to refineries in Szazhalmombatta, 30 miles south of Budapest.(126)

However, Hungary does have large deposits of uranium ore. Since 1956, the entire output of the uranium mines in the Mecsek Mountains has been sent to the USSR.(127) Hungary and the USSR signed an agreement in 1956 to build a 100 MWe reactor, but it was not fulfilled.(128) Whether this was because of the Hungarian revolution, or because of Soviet technical inadequacy is not known. On July 5, 1966, the Soviet Union reportedly signed a bilateral agreement to deliver two VVER-440 reactors to Hungary. One of these is under construction; the other is still on order. Both are scheduled to become operable in the 1980s.(129)

Romania

Romania is the only CMEA member with sufficient domestic sources of fuel to remain independent of the Soviet Union for all its energy needs. Romania has extensive petroleum and natural gas reserves, and was once a major exporter of crude oil. It still exports oil under previously-negotiated oil export agreements, and simultaneously imports oil from Iran, Saudi Arabia, Libya, Algeria, and Venezuela. Romania has the second largest natural gas deposits in Europe; in 1974, gas accounted for more than half of Romania's energy consumption. It has abundant coal resources of both lignite and anthracite. In conjunction with Yugoslavia, Romania has developed a hydropower dam, and it has a hydroelectric project underway with Bulgaria.(130) Finally, Romania possesses sufficient uranium for its own needs. About 1964, Romania unilaterially cancelled an agreement to supply uranium to the Soviet Union.

Romania has no operating commercial nuclear power reactors. It had planned to build a Soviet VVER-440 reactor that would go on line in 1983.(131) However, it is not clear whether this plan will be completed. For political reasons, Romania has attempted to purchase nuclear reactors from the West. One Romanian diplomat reported that his country has arranged to buy a natural uranium CANDU reactor from Canada. He emphasized that his country desired American, not Soviet, technology, and was willing to accept all safeguards and other nonproliferation measures; and he stressed the fact that Romania would participate in an international organization to manage spent nuclear fuel.(132)

Yugoslavia and Albania

Yugoslavia and Albania are the two East European countries that do not belong to CMEA. Yugoslavia has large quantities of coal, mainly lignite of poor quality; and this coal supplies about 68 percent of its energy needs. A large amount of oil is imported from the Middle East.(133) Westinghouse is constructing a 632 MWe pressurized water reactor in Yugoslavia which is expected to begin operation in 1979.(134) Feiveson and Taylor have reported that Yugoslavia also has a 615 MWe reactor on order from the Soviet Union.(135) Yugoslavia had a laboratory scale reprocessing facility at the Boris Kidric Institute, but that is now closed down.(136)

Albania is self-sufficient in energy, with extensive resources of oil as well as natural gas and lignite. China has given Albania technical assistance in the utilization of its hydropower resources.(137) Albania has, apparently, no plans to generate commercial nuclear power.

INTERNATIONAL MANAGEMENT OF SPENT NUCLEAR FUEL: SOVIET AND EAST EUROPEAN CONSIDERATIONS

In the early 1960s, during one of the first private talks about nonproliferation between a few representatives of the first four nuclear powers, our Soviet colleague clearly explained to us such a policy, saying in English: 'Nonproliferation is no problem, each one takes care of his own.'(138)

Bertrand Goldschmidt(138)

There are many proposals for international regimes to manage spent nuclear fuel on a regional basis, but the only model system in operation is in Eastern Europe. In principle, the system is simple: The Soviet Union exports only standard pressurized water reactors of the VVER series, supplies the enriched uranium fuel for them, and requires that the spent fuel be shipped back to the Soviet Union for storage; no spent fuel should remain in Eastern Europe. The European members of CMEA have all ratified the Nonproliferation Treaty. It has been claimed that this system provides a satisfactory model for the United States and the rest of the world to emulate.(139) I disagree.

Eastern Europe is a special case. The CMEA countries are politically and militarily under Soviet control; they are a captive market. Except for Romania, they can buy reactors only from the USSR. Enriched uranium fuel must come from – and be returned to – the Soviet Union. Bulgaria, for example, cannot let Canada, France, Great Britain, and the United States compete with the Soviet Union for reactor sales. Bulgaria must buy the VVER-440 reactor under Soviet terms. Emelyanov's remark to Goldschmidt that "each one takes care of his own" is misleading. In no other region of the world does one country so dominate her neighbors as does the USSR in Eastern Europe. However, even the Soviet system may be breaking down. Finland is currently constructing four nuclear reactors, two Soviet and two Swedish.(140) Spent fuel from the Soviet reactors must be returned to the USSR. What will be done with the fuel from the Swedish reactors? The USSR has sold power reactors to Cuba and Libya. Not until immediately before the contract was signed did Libya finally accede to Soviet demands to ratify the Nonproliferation Treaty. Cuba still has not ratified the NPT, although it has agreed to submit to IAEA safeguards. Unlike Finland, both Libya and Cuba are geographically too distant from Soviet borders to feel the undiluted pressure of the Red Army. Thus, it will be interesting to observe the actual disposition of spent fuel from reactors in Libya and Cuba.

Even in the special case of Eastern Europe, there is some evidence that the spent fuel storage regime does not work as well in fact as it does in theory: The Romanians are not required to ship spent fuel from their Westinghouse reactor to the USSR; nor is it known whether the Czechs promptly, or ever, return their spent fuel to the Soviet Union. As the CMEA countries attempt to increase their independence from Moscow, and as the Soviet Union begins to compete in world markets for sales of nuclear power reactors, Soviet control of spent fuel may not be as absolute as it now appears; fuel rods irradiated in nuclear reactors far from its borders will be difficult to regulate. Moreover, the Russians have decided that, regardless of the dangers of proliferation, their future energy needs – and those of their East European allies – will be met with breeder reactors. The USSR has, in fact, signed agreements with France, Japan, and the United States to conduct research on fast breeder reactors. In controlling spent fuel, Moscow may soon have trouble taking care ot its own.

The Soviet Union is attempting to develop nuclear power reactors into a major export industry: At Volga-Donsk, near Rostov, the Russians are creating a new city where "Atommash" will mass produce 1000 MWe reactors. Two CMEA organizations, Interatominstrument and Interatomenergo, are assisting in the development of this nuclear industry. The Soviet nuclear export market has excellent prospects. The Russians have sold their standard 440 MWe reactors to Cuba and Libya, and have approached the Philippines about possible reactor sales. However, after the reactors have been delivered, Moscow may be unable to enforce the terms of its contracts with non-contiguous countries outside the Soviet bloc.

The Soviet Union has supported the idea of regional efforts to inhibit nuclear weapons proliferation. On July 8, 1977, an article in Pravda asserted:

> At the request of a number of countries, the IAEA is at present studying the problem of creating regional centers for the nuclear power industry fuel cycle, as well as international "banks" – stores – for the plutonium obtained by the non-nuclear countries from power industry and experimental reactors. The creation of regional centers and plutonium banks would contribute to strengthening the nuclear weapons nonproliferation set-up and at the same time to the more efficient development of the nuclear power industry.(141)

No matter how strict a spent fuel storage regime is maintained in Eastern Europe, the USSR would probably not allow supervision or control of Soviet bloc spent fuel to pass to the IAEA or any other international body. Moscow is too sensitive about its uranium supplies, and too worried about its future sources of uranium for fueling breeder reactors to give up any sovereignty over uranium. Similarly, it is unlikely that that the USSR would create a fuel "bank" for its East European allies; if such a "bank" were created, the Hungarians, for example, could withdraw at will the energy equivalent or the plutonium equivalent of the spent fuel they had already deposited.

On the other hand, it costs the USSR nothing to support spent fuel storage regimes elsewhere. In general, the USSR is a strong advocate of nonproliferation measures. As an NPT-certified nuclear weapons state, the USSR could gain no advantage by further proliferation of nuclear weapons; it is sufficiently threatened by the United States, Britain, France, and China.

Indeed, the Soviet Union might wish to participate actively in international schemes for the management of spent fuel on which it has no prior claim. It is difficult to find storage sites for spent fuel in large countries; in small ones the problem is much more severe, and some East European countries are grateful to the USSR for removing radioactive fuel from their territory. Even the problem of decommissioning radioactive nuclear power plants is worrisome to many countries; in the American Congress there is opposition to accepting spent fuel from abroad. Thus, if the USSR were to accept spent fuel irradiated outside its borders, many nations would regard this as a gesture of good will. The potential economic value of spent nuclear fuel is great, and the USSR might easily decide to increase its possession of spent fuel by creating in its vast and underpopulated domain an international spent fuel storage depot for fuel irradiated outside of Eastern Europe. Indeed, the political and economic value of this depot could be so great that the USSR might, for the first time, be willing to submit to IAEA or other outside inspections.

International management of spent nuclear fuel is not a complete solution to the nuclear nonproliferation problem. It is not likely to work in Eastern Europe, but it is not necessary there because no CMEA country would attempt to build an atomic bomb. The Soviet Union could be expected to support international spent fuel storage schemes, so long as NATO members did not get too much credit or benefit. In fact, the Soviet Union might be so eager to participate in an international organization to manage spent nuclear fuel that it would open its borders to international inspectors. Because of this possibility for mutually-beneficial Soviet-American cooperation, vigorous efforts should be made to establish an international organization for the storage of spent nuclear fuel.

NOTES

(1) United States Arms Control and Disarmament Agency, *Documents on Disarmament, 1945-59* (Washington, D.C.: U.S. Government Printing Office) (hereinafter cited as *Documents on Disarmament*), document 2, p. 5.

(2) *Ibid.*, document 4, pp. 10-11.

(3) *Ibid.*, document 4, p. 11.

(4) *Ibid.*, document 5, pp. 17-24.

(5) *Ibid.*, document 27, p. 177.

(6) *Ibid.*, document 58, p. 262.

(7) *Ibid.*, document 30, pp. 185-86.

(8) Lincoln P. Bloomfield, Walter C. Clemens, Jr., and Franklyn Griffiths, *Khrushchev and the Arms Race* (Cambridge, Mass. and London: The MIT Press, 1966), p. 154.

(9) *Documents on Disarmament*, document 187, p. 737.

(10) Ibid., document 220, p. 879.

(11) Ibid., document 220, pp. 880-81.

(12) Ibid., document 225, p. 892.

(13) Ibid., document 306, pp. 1185-86.

(14) Ibid., document 252, pp. 978-80.

(15) Ibid., document 253, p. 981.

(16) Ibid., document 258, p. 998.

(17) Bloomfield, Clemens, and Griffiths, op. cit., p. 155.

(18) Jonathan D. Pollack, "China as a Nuclear Power," in William H. Overholt (ed.), Asia's Nuclear Future (Boulder: Westview Press, 1977), pp. 35-65.

(19) Pollack, op. cit.; Bloomfield, Clemens, and Griffiths, op. cit., pp. 123-30.

(20) Mao Tse-tung, "Talk with American Correspondent Anna Louise Strong," Selected Works of Mao Tse-tung, Vol. IV (Peking: Foreign Languages Press, 1961), p. 100.

(21) Pollack, op. cit.

(22) Ibid.

(23) "Statement by the Spokesman of the Chinese Government – A Comment on the Soviet Government's Statement of August 3 – August 15, 1963," Peking Review, VI, 33 (August 16, 1963), pp. 7-15, in William E. Griffith, The Sino-Soviet Rift (Cambridge, Mass: The MIT Press, 1964), p. 351.

(24) Ibid., p. 348.

(25) Ibid., p. 351.

(26) Ibid., pp. 344-45.

(27) Ibid., pp. 347-48.

(28) "Soviet Government Statement, August 21, 1963," Soviet News, No. 4885 (August 21, 1963), pp. 103-109, in Griffith, op cit., pp. 362-63.

(29) "Statement by the Spokesman of the Chinese Government – A Comment on the Soviet Government's Statement of August 21 –

September 1, 1963," Peking Review, VI, 36 (September 6, 1963), pp. 7-16, in Griffith, op. cit., p. 373.

(30) Ibid., pp. 374-75.

(31) Gerhard Wettig, "Soviet Policy on the Nonproliferation of Nuclear Weapons, 1966-1968," Orbis 12 (1969), pp. 1058-84; Benjamin S. Lambeth, "Nuclear Proliferation and Soviet Arms Control Policy," Orbis 14 (1970), pp. 298-325.

(32) Pravda, February 11, 1967, in Lambeth, op. cit., p. 313.

(33) George Quester, The Politics of Nuclear Proliferation (Baltimore: Johns Hopkins University Press, 1973); Toby Trister Gati, "Soviet Perspectives on Nuclear Nonproliferation," California Seminar on Arms Control and Foreign Policy, Discussion Paper No. 66, 1975.

(34) Quester, op. cit., p. 45.

(35) Ibid., p. 52.

(36) Gati, op. cit., p. 4.

(37) Theodore Shabad, "Soviet Union Steps Up Installation of Nuclear Power Plants," New York Times, January 14, 1977.

(38) Philip Hanson, "The Soviet Energy Balance," Nature, Vol. 261, May 6, 1976, pp. 3-5.

(39) A. M. Petros'yants, "Nuclear Power and its Significance in the USSR as a Source of Energy," in International Conference on Nuclear Power and its Fuel Cycle, Salzburg, Austria, 1975.

(40) David K. Willis, "Soviets Face a Different Kind of Energy Crunch," Christian Science Monitor, June 23, 1977: Hanson, op. cit.

(41) Sarah White, "Nuclear Power and the Five-Year Plan," New Scientist, April 21, 1977, p. 129.

(42) Hanson, op. cit.

(43) Shabad, op. cit.

(44) Ibid.

(45) Boris Belitzky, "Removing Radioactive Rubbish in the USSR," New Scientist, February 26, 1976, pp. 436-37; Boris Belitzky, "The Soviet Answer to Nuclear Waste," New Scientist, April 27, 1977, pp. 128-29.

(46) Zhores Medvedev, New Scientist, November 6, 1976; William E.

Farrell, "Ex-Soviet Scientist, Now in Israel, Tells of Nuclear Disaster," *New York Times*, December 9, 1976; Lloyd Timberlake, "Facts Still Scarce on Nuclear Disaster in Soviet Union," *Christian Science Monitor*, January 12, 1977; Zhores Medvedev, *New Scientist*, June 30, 1977, pp. 761-64.

(47) Simon Rippon, "Fast Reactor Progress in the Soviet Union," *New Scientist*, December 4, 1975, pp. 570-72.

(48) Vera Rich, "Environmental Protection under State Socialism," *Nature*, Vol. 259, February 12, 1976, pp. 438-39.

(49) Pyotr Kapitza, *Bulletin of the Soviet Academy of Sciences*, 1975.

(50) Elizabeth Pond, "Soviets Press for A-Power Leadership," *Christian Science Monitor*, August 25, 1976; Robert Toth, "Russ Try to Ease Doubts on A-Power," *Los Angeles Times*, May 30, 1976.

(51) Peter Stoler, "Soviets Go *Atomaya Energiya*," *Time*, October 30, 1978, pp. 68-71.

(52) "Soviet Says Nuclear Reactor is Supplying Power to Arctic Region," *New York Times*, October 21, 1973.

(53) Petros'yants, *op. cit*.

(54) *Ibid*.

(55) Robert Gillette, "Nuclear Power in the U.S.S.R.: American Visitors Find Surprises," *Science*, Vol. 173, September 10, 1971, pp. 1003-1006.

(56) Shabad, *op. cit*.

(57) Sarah White, "Soviet Expansionism," *New Scientist*, Vol. 70, June 24, 1976, p. 707.

(58) Petros'yants, *op. cit*.

(59) N. Samoilenko, "New Development: Energy Giant," *Izvestia*, March 4, 1978, in *Current Digest of the Soviet Press*, Vol. XXX, No. 9, March 4, 1978, p. 23.

(60) Shabad, *op. cit*.

(61) A. N. Grigor'yants, L. M. Boronin, I. A. Yegorov, and YE. P. Karyelin, "Development of the Nuclear Power Industry in the USSR," in *International Conference on Nuclear Power and its Fuel Cycle*, Salzburg, Austria, 1975.

(62) Petros'yants, *op. cit*.

(63) Rippon, op. cit.

(64) Petros'yants, op. cit.

(65) Theodore Shabad, "Soviet Operating Breeder Reactor," New York Times, July 17, 1973.

(66) Rippon, op. cit.

(67) Petros'yants, op. cit.

(68) Rippon, op. cit.

(69) Ibid.

(70) "Why the Russians Go All-Out for Nuclear Power," Business Week, August 2, 1976, pp. 52-53.

(71) Petros'yants, op. cit.

(72) O.D. Kazachkovskii, et al., "The Present Status of the Fast Breeder Program in the USSR," in International Conference on Nuclear Power and its Fuel Cycle, Salzburg, Austria, 1975.

(73) Business Week, op. cit.

(74) Ibid.

(75) "Atommash Takes Shape," Nuclear Engineering International, April, 1978, pp. 56-57.

(76) "East Germans Spur Atom Power with No Open Public Opposition," New York Times, May 30, 1978.

(77) Business Week, op. cit.

(78) Stoler, op. cit.

(79) V.S. Emelyanov, "Nuclear Reactors Will Spread," in C.F. Barnaby (ed.), Preventing the Spread of Nuclear weapons (New York: Humanities Press, 1969), pp. 65-71.

(80) A.P. Aleksandrov, "Atomic and Thermonuclear Power," Soviet Science, Vol. 45, No. 2, 1975, pp. 20-25.

(81) A. Panasenkov, et al., "Co-operation of the CMEA Member Countries in the Development of Different Reactor Types, Including Certain Aspects of their Nuclear Fuel Cycle, in International Conference on Nuclear Power and its Fuel Cycle, Salzberg, Austria,1975.

(82) Business Week, op. cit.

(83) K. Suetsugu, Industrial News Correspondent for the Japan Economic Journal, personal communication.

(84) Ibid.

(85) "Soviets Pushing Sales of Nuclear Plants," St. Louis Post Dispatch, October 8, 1978 (reprinted from the Washington Star).

(86) "Soviet Said to Offer Manila Atom Plant," Washington Post, February 15, 1978; "Soviets Offer Manila N-Plant," Boston Globe, February 15, 1978; Wall Street Journal, February 15, 1978.

(87) Daniel Yergin, "The Real Meaning of the Energy Crunch," New York Times Magazine, June 4, 1978, p. 99.

(88) Richard F. Staar, "Soviet Relations with East Europe," Current History, Vol. 74, No. 436, April, 1978, pp. 145-49, 184-85.

(89) Ibid.

(90) Dusko Doder, "Soviet Oil Price Increase to Eastern Europe Hinted," Washington Post, February 8, 1975.

(91) Peter Osnos, "Oil Crisis Aided Soviets but Poses Problems," Washington Post, February 8, 1975.

(92) Theodore Shabad, "West the Top Buyer of Soviet Oil in '76," New York Times, June 11, 1977.

(93) Staar, op. cit.

(94) Moscow Radio, September 20, 1976 and December 3, 1977; in Staar, op. cit.

(95) "On Scientific-Technical and Industrial Assistance of the Soviet Union to Other Governments in the Matter of the Development of Research on the Application of Atomic Energy for Peaceful Purposes," Izvestiya, January 18, 1955; see also Pravda, January 18, 1955.

(96) Jaroslav G. Polach, "Nuclear Power in East Europe," East Europe, Many 1968, pp. 3-12.

(97) United Nations, World Energy Supplies, Nos. 1-10; in Polach, op. cit.

(98) Jaroslav G. Polach, "Nuclear Energy in Czechoslovakia: A Study in Frustration," Orbis 12 (1968), pp. 831-51.

(99) Polach, "Nuclear Power in East Europe," op. cit.

(100) Pollack, op. cit.

(101) Gloria Duffy, "Soviet Nuclear Exports," International Security 3 (1978), pp. 83-111.

(102) Polach, "Nuclear Power in East Europe," op. cit.

(103) George W. Hoffman, "Energy Politics in Eastern Europe: Structural Changes in Production and Consumption and Resource Dependence," from: Proceedings: International Ex-Students' Conference on Energy, University of Texas at Austin, April, 1976, pp. 137-51.

(104) Panasenkov, et al., op. cit.

(105) Polach, "Nuclear Energy in Czechoslovakia," op. cit.

(106) Ibid.

(107) United Nations, World Energy Supplies 1960-1974 (New York, 1976), pp. 235, 106-111; in Hoffman, op. cit.

(108) Polach, "Nuclear Energy in Czechoslovakia," op. cit.

(109) Atomic Industrial Forum News Release, "AIF: Nuclear Powerplants Outside the United States as of May 1, 1976," June 2, 1976.

(110) Hoffman, op. cit.

(111) Ibid.

(112) Atomic Industrial Forum News Release, op. cit.

(113) Polach, "Nuclear Power in East Europe," op. cit.

(114) "East Germans Spur Atom Power with No Open Public Opposition," New York Times, May 30, 1978.

(115) Ibid.

(116) Ibid.

(117) Hoffman, op. cit.

(118) Atomic Industrial Forum News Release, op. cit.

(119) Dimo Velev, "Integration – A New Parameter of Friendship," Bulgaria Today, November, 1977, p. 15.

(120) *Ibid*.

(121) Hoffman, *op. cit*.

(122) Polach, "Nuclear Power in East Europe," *op. cit*.

(123) Atomic Industrial Forum News Release, *op. cit*.

(124) Hoffman, *op. cit*.

(125) David A. Andelman, "Hungary Looks Beyond Soviet for Its Oil," *New York Times*, May 15, 1978.

(126) *Ibid*.

(127) *Ibid*.

(128) Polach, "Nuclear Power in East Europe," *op. cit*.

(129) Atomic Industrial Forum News Release, *op. cit*.

(130) Hoffman, *op. cit*.

(131) Atomic Industrial Forum News Release, *op. cit*.

(132) Interview conducted by the author in the Embassy of the Socialist Republic of Rumania in Washington on February 16, 1978.

(133) Hoffman, *op. cit*.

(134) Atomic Industrial Forum News Release, *op. cit*.

(135) Harold A. Feiveson and Theodore B. Taylor, "Alternative Strategies for International Control of Nuclear Power," in *Nuclear Proliferation*, 1980s Project/Council on Foreign Relations (New York: McGraw-Hill, 1977), p. 173.

(136) SIPRI: Fuel Reprocessing Capabilities as of December 31, 1976, in Stockholm International Peace Research Institute, *World Armaments and Disarmament. SIPRI Yearbook 1977*, (Cambridge, Mass.: The MIT Press, 1977), p. 47.

(137) Hoffman, *op. cit*.

(138) Bertrand Goldschmidt, "A Historical Survey of Nonproliferation Policies," *International Security 2* (1977), pp. 69-87.

(139) Duffy, *op. cit*.

(140) Atomic Industrial Forum News Release, *op. cit*.

(141) *Pravda*, July 8, 1977.

3 Indian Ocean Basin

Onkar Marwah

INTRODUCTION

It is generally suspected that India used unsafeguarded spent reactor fuel, reprocessed into plutonium, to construct the nuclear device it exploded in 1974. Instituting controls over spent fuel has since become an important international issue, with the objective of preventing other states from following the Indian example.

Details vary, but all the current antiproliferation proposals require both international accounting and physical control of the spent fuel from different countries' nuclear reactors – especially those in LDCs. These proposals envisage spent fuel transfer to regional depositories, followed by storage and reprocessing under an international authority. Still to be resolved are technical questions about site selection, security and safety of transport and storage, apportionment of costs, and the national composition of storage facility overseers.

This chapter examines the feasibility of regional storage and reprocessing in the states bordering the Indian Ocean. It will suggest which countries might want to participate, what conditions will encourage their interest in an extranational facility, and the framework in which regional membership might be possible.

THE INDIAN OCEAN

The Indian Ocean is the third largest in the world. With its adjacent seas, it spans 72,000,000 square kilometers, or 14 percent of the earth's surface(See Map 1). Even before Western contact, the ocean was an important highway of commerce and interaction among the peoples of its littoral states. Since Vasco da Gama's sixteenth-century landing in southern India, European navies, commerce and imperial design dominated the ocean. Portuguese, Dutch, French and British naval squadrons and trading vessels competed against or fought with each other and the native peoples over spices, religion, and territory. By the mid-nineteenth century, the Indian Ocean had become a "British lake" and remained so until the 1940s. Today, thirty-five independent nations border the Indian Ocean.(1) All but three – South Africa, Israel and Australia – are third world states.

THE INDIAN OCEAN: CURRENT POLITICAL, ECONOMIC AND STRATEGIC CONCERNS

The world has suddenly awakened to the Indian Ocean's importance, for several reasons:

a) The origin off its shores, and the movement through its sea lanes, of 60 percent of the world's oil.

b) Great power disagreements over local wars now developing in a number of the African littoral states – Zimbabwe, Somalia, perhaps South Africa in the future.

c) Great power naval competition in the ocean.

d) The vast influx of conventional arms into the Persian Gulf states.

e) The apprehensions of littoral states that their rivalries might provoke great power military intervention in local affairs.

f) Anxieties with regard to the consequences of domestic turmoil in the important littoral states, e.g. Iran.

g) Concern over South Africa's nuclear plans.

This litany of troubles does not exactly presage instant success for proposals which, like the one for a spent fuel storage facility, demand high degrees of cooperation among Indian Ocean states as well as powerful states beyond the region. The path to regional spent fuel cooperation will probably have to weave deftly around the region's problems, rather than ignore them, if success is to be achieved.

MEMBERSHIP

Who should be included in an Indian Ocean facility? Technically, of course, any littoral state, from Australia at one end to South Africa at the other, could join an "Indian Ocean" storage regime. But practically, the membership list will be different. For obvious reasons, South Africa would refuse, and the non-European littoral states require its exclusion from any cooperative local venture.(2) Australia may wish to participate in a Southeast Asian rather than an Indian Ocean storage facility. That may also be true for Indonesia, Singapore and Malaysia. States skirting the Persian Gulf might contribute to an Indian Ocean facility, but they, too, might prefer a more local arrangement.

The Indian Ocean does not divide itself neatly for the purpose of siting a spent fuel storage facility. A universally acceptable location might be possible, but the membership list would have to be decided country by country.(3) Given the likelihood of separate Southeast Asian and Middle Eastern facilities, we suggest that all other states abutting the Indian Ocean – with the exception of South Africa – be considered candidates for an Indian Ocean spent fuel storage facility.

Of all the Indian Ocean candidate states, only India, Iran, and Pakistan have nuclear installations in place or under construction. The nuclear plans of the other littoral states – Bangladesh or Iraq, for instance-are too nebulous at present to cause concern about their responses to the projected storage regime. Practically, therefore, an "Indian Ocean" spent fuel storage facility will have to service only India, Iran, and Pakistan for the foreseeable future.

NUCLEAR FACILITIES IN INDIA, IRAN AND PAKISTAN

India

India possesses the following nuclear facilities:

1. Five research reactors.

2. Eight to ten power reactors of 250 MW each by the mid-1980s.

3. Two plutonium separation/reprocessing plants, able to handle 30 and 300 tons of fuel annually, respectively. (A third plant is under construction, and a fourth one is planned.)

4. A large scale prototype fast breeder reactor under construction for experiments with a thorium-fuelled breeding cycle.

5. Current research work in fusion and uranium enrichment technology.

6. Laboratory work in related areas, such as laser isotope-separation research.

7. Five heavy water plants in various stages of completion by 1980.

The Indian nuclear program – including its research capability, large scientific community, and substantial industrial base – is comprehensive enough to support civilian needs and, if required, weapons development simultaneously.

The country still lacks indigenous facilities for uranium enrichment and (temporarily) heavy water production, but the thrust of its nuclear program has been to increase self-reliance. India has neither signed the NPT nor agreed to a Pakistani proposal for a South Asian nuclear weapon-free zone, though it does not export its sensitive nuclear technology to other states. Formally, India describes itself as nuclear weapon-capable, not a nuclear-weapon power.

Recent attempts to assure that India is not, in fact, using its capabilities to increase its weapons potential, insist upon scrutiny of every nuclear installation in the country before more heavy water and enriched fuel are supplied to Indian power plants. India has not accepted these "full scope" safeguards, but its stand has not been completely rigid. The Indians say they will submit to such comprehensive safeguards if the super-powers proposing them accept the following counter-obligations:

1. An unconditional comprehensive test ban treaty signed by the United States, the Soviet Union, and Britain. (France and China could adhere to the ban if and when they choose.)

2. Freezing the superpowers' existing nuclear weapons systems and halting their future development.

3. Carefully planned and mandatory reductions in the superpowers' nuclear arsenals.

4. Agreement by the nuclear weapons states (France and China at their own volition) to safeguards on all their own nuclear installations, similar to those required of India.

These four conditions have become the basis of India's negotiating position in all nuclear matters.(4) They seem to preempt the country's accession to lesser agreements, such as participation in a regional spent fuel storage facility.

Iran

Iran operates one research reactor and has two 1200 MW nuclear powerplants under construction by West German firms. It has, however, recently withdrawn from a contract with France for the construction of two 900 MW reactors. The evidence is that, in the wake of internal political changes following the departure of the Shah, Iran's earlier ambitious nuclear power development plans are to be drastically cut back.

With or without its previous nuclear power plans, Iran would remain a nuclear novice for some time, with a modest scientific, industrial and technical structure. Given the U.S. Nuclear Nonproliferation Act of 1978, and nuclear supplier states' restrictions on exporting sensitive nuclear technology and materials, it would have been difficult for Iran to reduce appreciably the time required to develop general competency in nuclear research and development. While Iran may have, at some stage, considered building a national reprocessing facility, the London Suppliers' Group has proscribed the export of reprocessing equipment and technology to all other countries. France and Germany have also agreed to stop such future transfers.

Short of clandestine transfer or the "purchase" of expatriate scientific talent, the Iranians cannot acquire reprocessing technology from any external source. Iran will have to accept the numerous international restrictions on supply of sensitive nuclear technology if it wants to pursue an ambitious nuclear development program. The crucial question is whether nuclear supplier states, especially the European ones, will insist that Iran store its spent fuel at a regional or international facility, even at the risk of a retaliatory Iranian oil embargo. It is noteworthy that Iran is a financial partner with France, Italy, Spain, and Belgium in the Eurodif (under construction) and Coredif (planned) uranium enrichment plants. (Current reports suggest that Iran's new government may withdraw from the arrangement.)

Otherwise, Iran is a party to the NPT. The Shah had advocated a nuclear weapon-free zone for the Persian Gulf, and supported the Pakistani proposal for a South Asian nuclear weapon-free zone. There is no reason to presume that Iran's new leadership will modify those policies.

Pakistan

Pakistan's nuclear establishment, if modest compared to India's, is substantially superior to Iran's. A core of over 1,000 skilled technicians and scientists work at the Pakistan Atomic Center in Islamabad. The country possesses a 125 MW natural uranium commercial reactor at Karachi and had a 600 MW plant under construction at Kundian, which should be operating by 1982. Last year, the Pakistan government officially proposed the construction of twenty-four 600 MW nuclear power plants by the end of the century, in accordance with a 1975 IAEA study. Since 1960, the country has also possessed a five megawatt research reactor acquired under the American Atoms-for-Peace program. It is estimated that Pakistan can accumulate about 50 kilograms of separable fissile plutonium annually from its existing reactors. In 1976 a contract was initialled for a French-built plutonium reprocessing plant in Pakistan, but France has since withdrawn from the arrangement.

Pakistan has declared that it will not surrender its spent fuel to an international storage regime unless India follows suit. The country will probably not modify this resolve. In 1972 Pakistan rejected new, more restrictive Canadian safeguards on its Karachi power plant, even though Canada stopped supplying the fuel for it. Niger is now supplying the necessary natural uranium.

The reprocessing contract with France having been aborted, it can be assumed that Pakistan will accumulate its spent fuel locally until it figures out a national means for reprocessing plutonium. Such a course might impede Pakistani nuclear development plans, because the country would no longer qualify for further nuclear supplies or equipment from the London Suppliers Group of states. An independent path would arouse severe international pressure, despite Pakistan's growing indigenous competence in nuclear energy development.

Meanwhile, Pakistan does not intend to sign the NPT unless India signs first. It has also been the prime mover for a South Asian nuclear weapon-free zone.

WHO WOULD JOIN AN INDIAN OCEAN SPENT FUEL STORAGE FACILITY

The probable list of candidates for a regional storage facility (India, Iran, Pakistan) is much shorter than the possible list (all but South Africa).

Technical sanctions – denying nuclear technology and materials – which might succeed in compelling Iran and Pakistan to join a spent fuel facility, will fail with India, simply because it already has the most crucial elements of nuclear technology. Other types of sanctions, such as withholding aid or foodgrains, fall beyond the scope of this paper. They, too, would probably not succeed with either Iran or India, or probably Pakistan.(5)

Only under the most unlikely circumstance – a wide-ranging embargo against all three states – can one expect them to join an Indian Ocean spent fuel storage facility. Thus, an Indian Ocean Facility is improbable for the foreseeable future. If, by some freak of circumstance, it does come into being, it will have only two reluctant participants: Iran and Pakistan.

Locating an Indian Ocean Spent Fuel Storage Facility

Despite the unreality of presuming that Iran, Pakistan, and India could be persuaded to join a spent fuel storage facility, this exercise in considering that possibility should consider where it might be located and the international disagreements it might engender. It is extremely unlikely that Pakistan would accept a storage facility on Indian soil, or vice versa. One or the other would certainly reject an Iranian location. Therefore, the storage facility would have to be at a mutually agreed location outside of India, Iran or Pakistan, but within the Indian Ocean littoral states. The states open to site selection would be:

Region	Country	Remarks
East Africa:	Mozambique	(unsettled conditions)
	Tanzania	(possible but close to troubled area)
	Kenya	(")
Northeast Africa:	Somalia	(not an ideal location for the moment)
Arabian Peninsula:	North Yemen	(not an ideal location for the moment)
	South Yemen	(not an ideal location for the moment)
	Saudi Arabia	(Iran and India would not agree)
	United Arab Emirates	(")
Indian Subcontinent:	Nepal	(possible but the Himalayas are earthquake-prone)
	Bangladesh	(Pakistan would not agree)
Other:	Burma	(Burma would not agree)

The third alternative would be one of the Islands in the Indian Ocean:

Sri Lanka (independent)
Mauritius (")
Seychelles (")
Maldives (")
Malagasy (")
Comoros (")
British Indian Ocean Territory (Britain)
Diego Garcia (Britain, leased to U.S.)
Reunion (France)
Kerguelen (France)
Crozet (France)
Amsterdam (France)
Saint-Paul (France)
Prince Edward (South Africa)
Cocos (Australia)
Christmas (Australia)
Heard (Australia)
Mac Donald (Australia)
Laccadive-Minicoy-Amindivi (India)
Andaman-Nicobar (India)
Socotra (Democratic Republic of Yemen)
Jazirat Masirah (United Arab Emirates)
Jaza'ir Khuriya Muriya (United Arab Emirates)

Islands owned by Britain, France and Australia would be politically unacceptable to the three potential spent fuel contributors, which are strongly nationalist Asian states. The islands owned by South Africa would likewise be unacceptable. Pakistan and probably Iran would reject the Indian-owned islands. Islands owned by Yemen and the United Arab Emirates – apart from being located in a strategically turbulent area – are prone to political unpredictability. So are Zanzibar and Pemba.

Of the independent islands, Sri Lanka, Mauritius, Malagasy and the Comoros are beset with communal tensions which surface without warning from time to time. The Maldives and the Seychelles are relatively peaceful, but the former has a large population of Indian stock. The Seychelles are the famed "paradise islands" of Darwinian legend. In 1967, conservationists prevented Britain from building a 3600-meter air strip on one of the outer Seychelles islands; environmentalists the world over would object to their becoming a storage site for spent nuclear fuel.

Most of the islands listed actually comprise chains of islands (sometimes in the thousands), most of which remain uninhabited. The Maldives, for example, include 2000 islands grouped in some 20 atolls, or which only 200 are settled. But, the uninhabited islands are generally too small for major construction, and are frequently submerged at high tide. Those in the western Indian Ocean are also subject to fierce cyclones and monsoons. And it was recently discovered that three tectonic plates – Antarctic, African and Indian – join in the middle of the Indian Ocean. (See Map II) The ocean floor here spreads as much as ten centimeters a year as molten rock wells up along the fissure lines, accompanied by continuous earthquakes along the length of the ridge system. Potential earthquake epicenters dot the lines of contact of the three tectonic plates.

It is doubtful, though no impossible, that any of the newly-independent littoral states of the Indian Ocean would be willing to surrender a portion of their territory to international management.

It may instead be more feasible to select an island in the southern Indian Ocean for its political, environmental and seismic acceptability. Such an island could be detached from an existing group for extra-territorial jurisdiction and international management. Precedents for adopting this path exist. The British Indian Ocean Territory, for example, was created out of the Seychelles group.

The best candidate for hosting a spent fuel operation under international management are some French- and Australian-owned islands that lie along the imaginary divide between the southern Indian Ocean and Antarctica. They are uninhabited except for temporarily-licensed scientific expeditions, and have devolved to their proprietary states by accident rather than design. They are also seismically stable. Their forbidding climate and distance from population centers and sea lanes of commerce would frustrate terrorists and allow early warning of any attempted attack. U.N. jurisdiction over one of these islands would preempt the three Asian states' suspicions of each other, their neighbors, and the nuclear supplier states. The supplier states would probably find such a site the least objectionable.

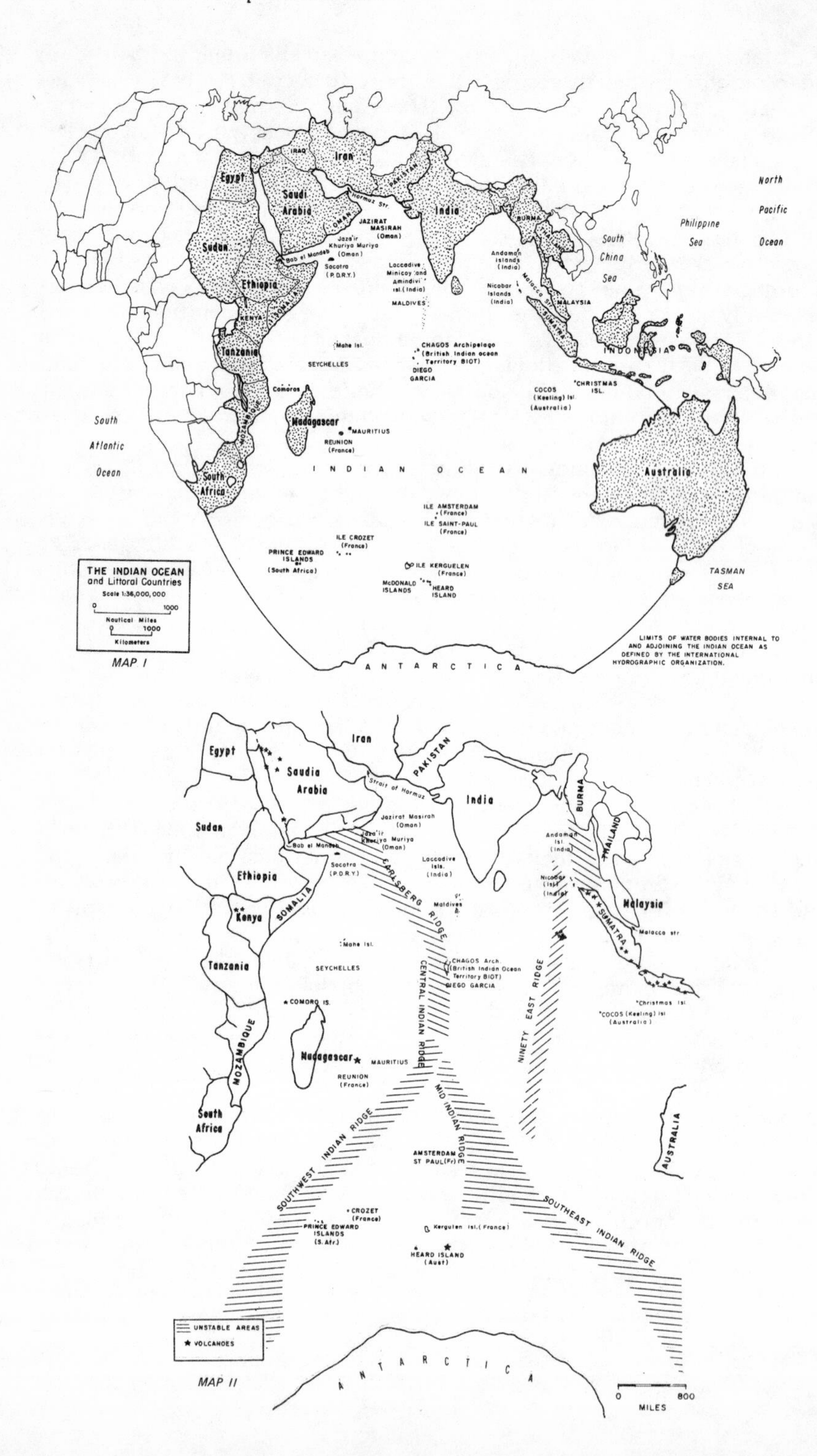
THE INDIAN OCEAN
and Littoral Countries
Scale 1:36,000,000
Nautical Miles
Kilometers
LIMITS OF WATER BODIES INTERNAL TO AND ADJOINING THE INDIAN OCEAN AS DEFINED BY THE INTERNATIONAL HYDROGRAPHIC ORGANIZATION.
INDIAN OCEAN
ANTARCTICA
UNSTABLE AREAS
VOLCANOES
MILES

MAP I

MAP II

Some of the consequences that would follow the arrangement as suggested can be stated. Storage or return of the spent fuel to the supplier state would not occur. Multilateral or international management of the storage complex would be required. If the spent fuel is to be used, then reprocessing facilities would have to be set up locally and be open to the oversight of all the managers. Both markets and the relevant overseer agencies would be distant from the storage and reprocessing site. Transporting the spent or reprocessed fuel would be an arduous task over long distances.

It is possible that some would view the preceding as shortfalls in arrangement. Perhaps it would be useful to see them in trade-off patterns with what is gained and what is lost in their implementation. If it is a worthy cause to initiate local participation in the principle of international accounting of spent fuel without having to impose politically vexing sanctions, then the costs may not be unreasonable.

Thus, there is no inherent value in requiring that the spent fuel be returned to the territory of the supplier state. Multilateral or international management, by itself, is neither an original concept nor self-defeating in this particular instance against the fear that sensitive nuclear technology would thereby be disseminated. Of the three presumed participants in an Indian Ocean storage facility only India with its breeder reactor program can presently make the economic case for reprocessing technology. As it has already developed the same for its own purpose the question of divulging anything new does not arise.

The Pakistani and any future Iranian case for reprocessing rest on projected nuclear power development. In the interim an economic case for reprocessed fuel in place of otherwise readily-available low enriched commercial fuel cannot be made easily – especially if an international nuclear fuel bank guaranteeing supply at reasonable rates is set up. The point is that an Indian Ocean spent fuel storage facility would not require the simultaneous construction of both storage and reprocessing facilities. The second function could be deferred to a later date. During the intervening period the facility could remain a spent fuel storage and nuclear waste management complex.

Finally, proximity to overseer agencies and markets, or the problems of transportation are less substantive issues, particularly when in-place multilateral oversight is accepted and transport initially requires a one-way arrangement.

COST, SECURITY, SAFETY AND CONTROL

Detailed expenditure calculations lie beyond the scope of this paper, but the issue is raised to seek answers to the following questions: (a) who pays?; and (b) would the storage facility be worth the cost incurred?

To make the proposal palatable it would be worthwhile to conceive initial capital expenditures being provided by the supplier states. Later on, user-charge modalities could be negotiated with the participating states. As the principle for having such a facility and the learning process which accompanies participation become widely shared there is no reason to assume that user-states would decline to share the monetary burden.

The facility's cost benefit ratio cannot be figured solely in monetary terms. With the relatively modest deliveries of spent fuel from India, Iran and Pakistan, the storage facility would most likely lose money. But there are compensating political benefits in institutionalizing a regional spent fuel storage regime.

An eventual monetary off-set for the southern Indian Ocean (island) storage facility can be conjectured if its depository role is subsequently (or simultaneously) expanded to cover spent fuel from adjacent regions. Indeed, Latin American, Southeast Asian, Middle Eastern and landlocked African countries are as "close" to the islands in question as the Asian participants being reviewed. Their participation would probably make the difference in financial terms.

The problems of adequate security arrangements are organizational. Those of safety against hazardous discharge of the stored nuclear fuel into the environment are technical in nature. If the facility were internationally managed, it could be policed by a U.N. security force. The technical questions with regard to poisonous seepages into the subsoil still await an optimal solution. In any case, the danger from accident would be less were the material deposited at a place remote from human habitation.

DIFFERENTIATED SPENT FUEL STORAGE REGIME

An alternative to the desired goal (full participation) and the probable situation (no participation) can be suggested in the form of a "differentiated storage regime." Institutional effort in this instance would be sensitive to the existing imbalances in the nuclear development, capability and technological proficiency of the states in review. The latters' intentions, paper plans and proclaimed hopes would be tacitly ignored until such time as they are implemented. The overriding policy objective here would be to gain legitimacy for the principle of an international accounting of spent fuel.

The process could begin by inviting the three states to deliver all their accumulated or future spent fuel from presently safeguarded reactors into the regional depository. None of them should object to such a request as it merely activates, with a minor modification, their legal obligation to return the spent fuel to the supplier states. This would immediately prevent, without any fanfare, the potential diversion of spent fuel from four Indian, two Iranian (under construction), and two Pakistani (one under construction) power reactors. The arrangement would also include one research reactor each in Iran and Pakistan. All future reactors constructed under international aegis in any of the three states would similarly be required to transfer the spent fuel to the Indian Ocean depository.

The obvious problem with this approach is that a number of Indian nuclear facilities (and future indigenously-constructed projects) would be excluded from fullscope oversight. In formulating policy, the issue in reference to that exclusion can be stated as follows: is it preferable to seek limited Indian accession to a process that legitimizes an international spent fuel regime, or is it better to delay application in an attempt to gain India's full accession but risk losing the assent of all three states? Responses to the question can affirm either the need for equity or that of pragmatism. They cannot, it would seem, reflect both those preferences yet secure implementation of a spent fuel storage regime. The situation is that the Indians have already escaped the enclosure in terms of the requisite nuclear techniques, materials and fabrication capability. Their information and experience in those areas can neither be rescinded nor unlearnt.

As far as anxieties exist with regard to India's potential for nuclear weapons, the strong probability is that the weapons could be manufactured even if all nuclear installations in the country came under fullscope safeguards. Those who suspect Indian intentions in this area could not be assuaged because fullscope safeguards were applied to India's known nuclear facilities. Indeed, the "suspicion" argument would cut both ways: the Indians, aware of others' doubts about their own nuclear plans, could not but assume adversary effort to the same end. As such, it is extremely unlikely that the Indian "weapons option" could now be surrendered. The fact is that, in losing their nuclear virginity, the Indians might have savored the exhilaration of the moment but – like others before them – must live with the consequences of the act.

There are other mitigating realities in relation to dire predictions about Indian intent. A nuclear weapons option is not an operational weapons system. Unlike the arguments in behalf of the Israelis, a bomb in the basement or on the shelf is not a viable premise for Indian strategic assessments. A few nuclear weapons would be meaningless against militarily-inferior subcontinental neighbors and useless against militarily-superior states beyond South Asia. So, an actual decision to proceed with the development of a nuclear weapon strike force is minimally what would make India into a nuclear weapon power. No such decision can be ascribed to the Indian government. Trigger lists of items and activities which lead toward the acquisition of a nuclear strike force are maintained by the great powers and any Indian effort in that endeavor would be quickly revealed. Until such activity can actually be discerned it may not be unreasonable to accept the contention that India is a nuclear-weapon-capable but not a nuclear-weapon power. If this line of thinking can be sustained, then the urgency of requiring full Indian acquiescence to an international spent fuel storage regime is somewhat reduced.

In the event, Indian statements about their low nuclear weapon proclivities – however skeptical they appear – could be accepted for pursuing the greater goal of establishing an international spent fuel storage regime. Meanwhile, through sustained bilateral contacts, the Indians could be encouraged to maintain their non-nuclear-weapon status, desist from further nuclear explosions, and voluntarily increase participation in the spent fuel storage facility.

The strategy described in this subsection of the study to legitimize an Indian Ocean spent fuel storage regime is applicable to other regions with inconvenient states such as India, e.g., Latin America, the Middle East, the Pacific. While the specific conditions would differ, the same approach of differentiating among states (in reality, accepting differences) according to local nuclear competencies could be applied.

CONCLUSION

It is no longer possible to divide the world neatly between states with nuclear weapons and states without. India's recent history shows that some states can sit on their weapons fabrication capabilities without using them to the maximum. The differentiated fuel management system proposed here suits this new nuclear environment. It is more likely to win acceptance in the Indian Ocean region than a "purer" system for combating proliferation. It may well serve as a model for other regions, since differing levels of nuclear development are common in all areas, and are often ignored in planning international fuel management schemes. And most important, a differentiated scheme will not become superfluous. Even if some of its members achieve the capability to build atomic weapons, it is an international arrangement capable of absorbing power shifts among nations: imperfect but practical, a modest force for improving a seriously flawed environment.

NOTES

(1) These are South Africa, Mozambique, Tanzania, Kenya, Somalia, Ethiopia, Sudan, Egypt, Afars & Issas, Israel, Jordan, Saudi Arabia, Yemen Arab Republic, Peoples Democratic Republic of Yemen, Oman, United Arab Emirates, Kuwait, Iraq, Iran, Pakistan, India, Bangladesh, Burma, Thailand, Malaysia, Singapore, Indonesia, Australia, Sri Lanka, Malagasy, Seychelles, Mauritius, Maldives, Comoros.

(2) Under certain circumstances, South African participation would be possible, e.g., a managerial division in an otherwise single spent fuel regime authority to negotiate separately with South Africa.

(3) For obvious reasons, states such as Egypt and Israel have been left out of the arrangements considered in this study.

(4) The Indian 'package' proposals, with an emphasis in their interdependence, are not new. They were first stated at the International Atomic Energy Conference in 1956, have been frequently repeated since then – but have acquired a new urgency in the post-Indian-nuclear-explosion stage.

(5) Iran's financial solvency based on oil is well-known. Less well-known is the fact that India has accumulated foodgrain reserves of 20 million tons and foreign exchange holdings of $5 billion (increasing @ of $2

billion annually as of last year). Pakistan's position is less strong but it can count upon the financial support of friendly Middle Eastern states.

(6) A problem exists in that, to avoid pollution, the proprietory states may have laws that prohibit the use of the islands for activities involving poisonous chemicals. A solution would lie in persuading the concerned proprietory state to enact and enforce legislation similar to Canada's Arctic Waters Pollution Act of 1970. Under the latter, the Canadian government is empowered to precisely control the exploitation, transportation, waste disposal, location, methods of operation, hardware, and insurance relating to its North Slope oil.

4 Latin America

Victoria Johnson
Carlos Astiz

INTRODUCTION

Little if any consensus now exists among the nations of Latin America about the need for regional cooperation in the storage of spent nuclear fuel. The countries with substantial nuclear programs – Mexico, Argentina, and Brazil – face no imminent storage problem, and they disagree about the importance of nonproliferation and the proper uses of nuclear technology. The other nations of the region are more concerned with translating their nuclear development plans into reality than with solving spent fuel storage problems that have remote significance. Many Latin American countries believe that the nations with a more urgent need for interim storage of spent reactor fuel will establish a system for themselves, leaving Latin America free to choose whether it wishes to join the established regime.

Uneven levels of nuclear development and disparate views about the importance of international oversight for spent fuel make any discussion of Latin views on cooperative storage facilities speculative. However, the possible shape of Latin participation in a regional fuel cycle management facility can be determined by examining existing nuclear programs in Latin America, Latin attitudes towards nonproliferation, and the political aspects of spent fuel storage arrangements.

ARGENTINA

Argentina's nuclear program was designed to minimize reliance upon foreign suppliers, and to maximize cooperation with foreign expertise. A natural uranium technology is central to this objective. In keeping with its basic goal to possess an autonomous nuclear program, Argentina is the only country in Latin America to have both a fuel fabrication facility and a plutonium separation plant. In addition, an experimental heavy water plant is under construction; bids were opened in August

1978, and the plant is scheduled to go into operation in 1981. The construction of an industrial heavy water plant capable of producing no less than 250 tons per year is also under consideration.

The Argentine natural uranium nuclear energy program is managed by very well-trained specialists, most of whom have received graduate training in the United States and Europe; they are employed by the National Commission of Atomic Energy (CNEA), which awards at least 20 fellowships a year to members of its professional staff so they may pursue Ph. D. training abroad. Additional personnel receive specialized training to handle specific projects; approximately 55 individuals were sent to Italy and Canada during 1978 to increase their knowledge of the Embalse nuclear power plant.

The first nuclear reactor in Latin America was Argentina's Atucha I, a 319 MW natural uranium reactor provided by West Germany's Siemens, A.G. Located 70 miles northwest of Buenos Aires, this reactor commenced operation in 1974. Argentina's second natural uranium power plant is currently under construction at Embalse in the province of Cordoba. It was reported to be 60 percent complete at the end of 1977, and will generate 600 MW after its scheduled start-up in 1982. The reactor vendor, Atomic Energy of Canada, Ltd., is also helping Argentina develop permanent fuel fabrication facilities, scheduled for operation in 1979. Until then, fuel elements for Atucha I are being provided by Kraftwerk Union, A.G., the nuclear arm of Siemens.(1)

A third nuclear plant, Atucha II, was ordered from Atomic Energy of Canada in 1976. Its reactor was to have been similar to the one under construction at Embalse, but Atucha II has been delayed by a Canadian demand for additional reactor safeguards. Argentina, reluctant to accept the additional safeguards, is now looking for other reactor suppliers.(2)

The Argentine nuclear power program relies on domestic natural uranium. Reserves have been estimated at 60,000 metric tons, which can be mined at a cost of less than $30 per pound of uranium oxide.(3) The country operates its own facilities to extract and refine the uranium ore into usable fuel. The five research reactors depend on imports of highly enriched uranium from the U.S. For the present, heavy water must also be bought from the U.S., although an experimental heavy water plant should begin operation in 1981. Its output will be adequate to replace existing stocks, but the first load of heavy water for additional reactor will have to be imported.(4)

Argentina's small reprocessing plant started producing in 1968 and can accept 200 metric tons of spent fuel per year. Plans to build a large reprocessing plant have been under consideration for some time. Available data indicate that the Atucha I power reactor produces 70 kilograms of plutonium per year; thus Argentina should have accumulated no less than 210 Kg of plutonium, and possibly 250 Kg, since Atucha I started operating in 1974. Because published figures vary widely, it is not possible to estimate accurately how much plutonium is available from the five research reactors. Whatever the exact amount, it is clear that Argentina has enough plutonium to produce nuclear explosives and the trained manpower to construct them. Despite an obvious technical capability to do so, Argentina has not stated any intention to exploit a nuclear weapons option. Actual efforts to produce nuclear weapons have been neither publicly announced nor detected by the IAEA, which safeguards the plutonium produced by the Argentine nuclear program.

The capability to exploit a nuclear weapons option, and the refusal to sign any international agreements that would prohibit it from building nuclear weapons suggest that Argentina might sometime seek a nuclear-weapons capability. Also, Brazil's ambitious nuclear program may conceivably prompt an Argentine response at least to develop a rudimentary nuclear capability.

> The use of her (Argentina's) temporary technological superiority to provide a nuclear veto power over any totally unpalatable Brazilian action may be the only way that Argentina can maintain a credible and independent position within the hemisphere.(5)

In comparison to Brazil, Argentina has suffered economic stagnation and political crisis in the last few years. The country must have economic growth and political stability if its leadership hopes to maintain parity with Brazil's influence within Latin America. Without this, the option to exercise even a simplistic nuclear option may become more attractive. Diplomatic measures designed to trade-off a non-proliferation commitment or the ratification of the Treaty of Tlatelolco, a Latin American compact banning nuclear weapons in the region and establishing it as a nuclear-weapons-free zone, for political support and a reduction of external pressures toward greater respect for human rights are likely to serve short-term objectives. In the long run, Argentine nuclear pursuits will probably be determined by Argentina's capacity to match Brazil's influence in other arenas. The shift now taking place in Latin America's traditional balance of power does not favor Argentina.(6)

A senior Argentine military officer recently suggested how nuclear weapons might tempt a security-conscious leadership:

> a) In the political arena: ...to guarantee the total autonomy of the state in this matter, in order to reaffirm its geopolitical role in the space which historically belongs to it. . .
>
> b) In the national security arena: to reach, in no more than ten years, a situation of pure and undisputed autonomy in the areas of nuclear activities and subsequent use of its potential capabilities, as well as to maintain the subregional balance on this matter.(7)

To the extent that the Argentine leadership perceives a multinational storage facility solely as an incentive toward nonproliferation, active Argentine participation is difficult to predict. As additional power plants start producing spent fuel and the difficulties of managing it become apparent, it is possible that the leadership might consider it advantageous to transfer at least some spent fuel to a multinational management facility over which it could exercise some supervision. This option might become more attractive if other Latin countries with nuclear reactors were to join some multinational arrangement.

BOLIVIA

Although Bolivia has had a Nuclear Energy Commission for some time, its relatively low energy needs and its adequate oil and gas reserves – it exports both – make nuclear energy unnecessary. In April 1978, however, Bolivia signed a nuclear cooperation agreement with Argentina, the details of which have not been publicly released.(8) Quite likely, the agreement is a product of Argentina's desire to compete with Brazil for the good will and mineral resources of Bolivia, and of Bolivia's desire to keep up with its neighbors' interest in nuclear technology. Without a current nuclear program, Bolivia has no direct interest in the issue of spent fuel storage. Bolivia has ratified both the Nuclear Nonproliferation Treaty and the Treaty of Tlatlelolco.

BRAZIL

Through the Atoms for Peace program, Brazil obtained its first research reactor, located at the Sao Paulo Atomic Energy Institute. Since 1960, additional research reactors have been located at the Belo Horizonte Institute of Radioactive Studies and the Rio de Janeiro Institute of Radioactive Studies, and, since 1965, at the Rio de Janeiro Institute of Nuclear Engineering. Brazil also has three subcritical reactors in Belo Horizonte, Recife, and Sao Jose dos Campos.

Brazil has signed bilateral agreements for scientific and technical cooperation in nuclear development with Bolivia, Canada, Chile, France, India, Israel, Italy, Paraguay, Peru, Portugal, Switzerland, The United States, and West Germany. A 1967 agreement with France established terms for joint research in thorium technology, uranium exploration, and research and power reactor construction . Franco-Brazilian nuclear cooperation also includes a contract, signed in July 1975, for the development of a fast breeder reactor. This project Cobra (Cooperation Brazil) calls for the French company Technicatome to build a breeder reactor similar to the French Rhapsodie; commercial operation should start between 1985-90. A 1974 agreement with an American firm, Gulf General Atomic Company, outlined Brazil's plans for cooperative development of a gas-cooled fast breeder reactor employing thorium fuel.(9)

Brazilian nuclear cooperation with Germany was established through agreements in 1953, 1969, and 1975. The 1969 agreement provided for scientific and technical cooperation in developing nuclear hardware and in training Brazilian scientists. In June 1975, Brazil agreed to purchase a complete nuclear fuel cycle system from West Germany. The contract covers uranium exploration and the construction of facilities for uranium enrichment, fuel fabrication, and chemical reprocessing for plutonium extraction. The agreement also calls for the construction of eight 1,250 MW reactors by 1990. These would supplement the original 626 MW Angra-1 reactor; located 60 miles southwest of Rio de Janeiro, this is now being constructed by Westinghouse and is scheduled to begin operation in 1980.

Luiz Claudio de Almeida Maghales, President of Furnas Centrais Electricas, S.A., a government-operated electricity company, optimistically projects that by the year 2000 Brazil will have 63 nuclear reactors with a total generating capacity of 81,000 MW. Hypothetically, the technology Brazil acquires from its agreement with West Germany could enable it to build the 54 reactors needed to realize Maghales' prediction. Financially, Maghales' projection is unrealistic, but it does emphasize that Brazil is accelerating its nuclear power program.(10)

Two significant complications have delayed Brazilian nuclear development. The anticipated completion date for its first power reactor, the Westinghouse-built Angra-I, has been delayed several times because of repeated fires (reference appears in Jornal do Brasil, Rio de Janeiro, Nov. 1, 1977.); a major one in 1977 caused an estimated $10 million damage. Recent statements in the Brazilian press imply that there have been some 70 small fires at this reactor site, a number suggesting arson. If sabotage has caused the rash of fires, then the reactor site's physical security is clearly inadequate. Angra-I's problems have significance for Brazil's entire nuclear power program because tentative Angra-2 and Angra-3 reactor sites are located nearby. The potential for multiple reactor damage, as well as atmospheric and human contamination could be considerable. Furthermore, if Brazil cannot protect its domestic nuclear facilities from sabotage, it might best store its spent fuel elsewhere.

The second complication stems from Brazil's agreement with West Germany. When the Bonn-Brasilia sales contract was signed in June 1975, the Germans expected to supply enriched fuel to Brazil from the enrichment plant at Almelo in eastern Holland that is operated by the Anglo-Dutch-German URENCO consortium. The Dutch Parliament, however, refused to approve the necessary enlargement of the plant unless Brazil agreed to store its unseparated spent fuel and forego reprocessing. Brazil rejected the Dutch proposal, but expects Germany to fulfill all terms of the original contract. In April 1978, the Germans accelerated construction of an enrichment plant located at Gronau near the Dutch border. The British are apparently assisting Germany's attempt to assure enriched uranium for Brazil's reactors. Gronau could supply fuel to Brazil by 1983 if there are no delays in construction and startup.(11) It is difficult to assess the full ramifications of the recent delays in executing the terms of the agreement. Uranium mining in Brazil is proceeding slowly and German equipment transfers for the first reactor have been nonexistent. Former projections that the new German reactors would be operating in Brazil by 1990 must obviously be revised. Also, because costs for nuclear equipment have risen, Brazil may decide to purchase only six German reactors instead of eight.

Despite the safeguards on the German transfer of nuclear equipment, Brazil remains committed to developing an autonomous nuclear program. It has not signed the NPT on the grounds that this agreement endorses continued nuclear developments in the U.S., France, China, The USSR, and Great Britain, while restricting them in all other nations. Actually, the NPT and the Treaty of Tlatelolco seek to restrict military applications of nuclear technology, not civilian use of nuclear power.

CHILE

Chile's nuclear program got under way in 1969 with the purchase of a 5 MW research reactor from a British firm, Fairly Engineering Ltd. This unit operates with enriched uranium provided by the U.S. A second nuclear reactor became operative in October 1977; located in Lo Aguirre, it has a power of 20 MW. It is reported to have been designed by the army's nuclear engineers, with Spanish technical assistance.(12) Chile's National Power Authority has conducted a feasibility study for a nuclear power plant in the 600 to 800 MW range. The plant, to be located near the northern city of Antofagasta, would desalinize sea water as well as generate electricity.

Chile's nuclear options are limited. Although some domestic uranium deposits have been located and prospecting is under way, estimates of uranium reserves vary widely. Furthermore, human, technological, and financial resources are in short supply, a combination which makes the launching of a major nuclear power program unlikely in the near future.(13)

Chile has ratified neither the NPT nor the Treaty of Tlatelolco. The country's current position on nonproliferation appears to be determined more by Argentina's non-participation and the desire to emphasize its political autonomy than by any realistic military plan to build nuclear weapons.

COLOMBIA

Colombian nuclear activities are confined to research. With surplus hydro-electric capacity, accessible coal, and some petroleum, Colombia has devoted little effort to nuclear development.

A two kilowatt research reactor is managed under the Institute for Nuclear Studies, the national agency in charge of nuclear concerns. Although there are some indications that Colombia may purchase a nuclear power plant within the next 10 years, no concrete plans yet exist.

Colombia is a full party to the Treaty of Tlatelolco but not to the NPT, reflecting political opposition to the nonproliferation policies of the nuclear powers rather than an intention to construct nuclear explosives.

CUBA

The Soviet Union constructed and supplies fuel for Cuba's only research reactor; this operates without International Atomic Energy Agency (IAEA) safeguards. A 1974 Cuban-Soviet agreement on cooperation for technical development provided for the construction of two 400 to 500 MW power reactors by 1980. In addition, Cuba plans to build nine medium-sized reactors before the year 2000.

Although Cuban leaders publicly endorse regional denuclearization efforts, Cuba has not signed either the Treaty of Tlatelolco or the NPT. Cuban participation in a cooperative spent fuel venture seems unlikely because Soviet-supplied uranium fuel must be returned to the Soviet Union. Modifications of this standard Soviet policy which would allow Cuban spent fuel to go elsewhere would require Soviet consent.(14)

MEXICO

The National Energy Institue administers nuclear development in Mexico. Its program differs from Brazil's and Argentina's in two major ways: It is guided exclusively by economic development objectives because its leadership has firmly renounced the development of nuclear weapons; and it has always been under civilian control.

Nuclear power plant installation is the responsibility of the Federal Power Commission, an agency of the national government which monopolizes the commercial generation and distribution of electricity. It is now building its first power plant at Laguna Verde; and at least one – and probably two – more are slated for construction on the same site. Laguna Verde I and II will have enriched uranium light-water reactors supplied by General Electric; each will generate 654 MW.(15) The United States will supply enriched uranium for these and other plants now being planned, subject to IAEA safeguards.

Two Mexican research institutions also operate reactors: the Salazar Nuclear Energy Center owns the one megawatt Triga-MK III; and the National University of Mexico owns the SIR-100 zero power reactor, which it purchased from West Germany.

A small fuel fabrication facility has been built in Chihuahua, where a pilot reprocessing plant is under construction. Mexican uranium reserves appear to be limited, with proven stocks of 6,000 metric tons. Because the export of uranium ore is prohibited by law, the Mexican Government hopes that intense exploration will unearth enough uranium to satisfy its nuclear development plans.

To a certain extent, nuclear development in Mexico came about as the result of recommendations made by the Stanford Research Institute in a 1968 report. The Institute concluded that Nuclear power should be developed in Mexico because alternative energy sources were limited. The assumptions of the report, however, have been made obsolete by recent discoveries of sizeable oil and natural gas reserves, and by the marked increase in fossil fuel prices. These reserves may not only meet Mexico's growing energy demands, but may transform the country into a significant energy exporter. Nevertheless, the shortage of fresh water still makes nuclear power attractive for desalinization.

The Mexican ruling elite appears to have no interest in nuclear weapons; it values nuclear energy as a stimulant to economic development. Mexico's policy of maintaining a weak military establishment, and its close ties with the United States make it unlikely that improved nuclear capabilities will alter its peaceful intentions.

As the major Latin American nation clearly committed to nonproliferation, Mexico has ratified the NPT and sponsored the Treaty of Tlatelolco. This policy enjoys the support of those parties and groups that are permitted to articulate their views in the Mexican political system. Nearly a decade ago, the government's policy of denuclearizing Latin America received the endorsement of the tolerated opposition on both the right and the left. Succeeding administrations have reiterated their commitment to regional and global nonproliferation. Because rejection of the nuclear weapons option has been a cornerstone of Mexican national policy, participation in multinational spent fuel management seems likely.

PANAMA

Panama has no nuclear program. The United States, however, maintains a power reactor in the Panama Canal Zone. There are no known plans for research or power reactor construction. Panama is full party to the Treaty of Tlatelolco, but not to the NPT.

PARAGUAY

Paraguay has one subcritical nuclear unit which it purchased from Brazil. No significant nuclear facilities currently exist, although Brazil recently announced plans to construct a nuclear research center outside of Asuncion to explore medical, agricultural, and industrial applications of nuclear technology. Paraguay is full party to the NPT and the Treaty of Tlatelolco.

PERU

Peru has signed the NPT and the Treaty of Tlatelolco. Although the country has some hydroelectric potential in the Amazon region and proven oil reserves of 3,500 million barrels(16), it has recently shown increased interest in nuclear technology. In March 1977, Peru concluded an agreement with Argentina for the construction of a zero-power reactor for Peru's National Center for Atomic Energy. This was scheduled to begin operation in late 1978 and will be used by the Peruvian Institute of Nuclear Energy to train its personnel.(17)

In November 1977, these two countries signed another agreement; it calls for the Argentine Atomic Energy Commission to supply a 10 MW research reactor and other equipment that will make up the core of Peru's Nuclear Research Center. Partial financing is to be provided by Argentina's National Development Bank.(18) Investments of this nature, at a time when Peru is having extreme difficulty in both international credit markets and domestic finances, imply a major commitment to nuclear research.(19)

VENEZUELA

Venezuela possesses not only impressive petroleum resources but considerable hydroelectric potential. Consequently, Venezuelan nuclear development has never progressed beyond a 3 MW research reactor which started operation in 1962. However, because insufficient staffing and projects have kept it closed for years, this reactor has been of little value. Any current research is performed with fuel supplied by the U.S. under IAEA safeguards.(20)

Venezuela wants to diversity its energy base. In March 1978, Brazil and Venezuela signed a $2 billion agreement to construct a 9 million kilowatt dam in Venezuela's Guyana region. This complex, scheduled for completion in 1985, will be the third largest in the world.

Interest exists, too, in nuclear energy investment. Initial negotiations for the purchase of a reactor have taken place, and will probably be consummated before the year 2000.(21) Venezuela possesses significant thorium and uranium reserves. The bulk of the uranium deposits lie in the Orinoco River basin, which has been claimed by Guyana. Venezuela will probably conform to the Mexican policy of prohibiting foreign extraction of uranium ore, but it is likely to consult with foreign firms about uranium exploration and extraction.

Although its official position on an international facility has not been articulated, Venezuela might be willing to cooperate in a regional fuel cycle facility. It is a full party to the NPT and the Treaty of Tlatelolco and seems to be a potential participant in a regional fuel cycle or storage enterprise. In fact, Venezuela's endorsement of nonproliferation, its opposition to nuclear weapons in Latin America, and its promotion of regional solutions are so complete that it might even be willing to host a spent fuel management facility.

REGIONAL CONSIDERATIONS

Significant differences exist in the sophistication of Latin American nuclear programs. Each nation's approach to nuclear development has depended heavily upon political implications as well as resource endowments. Argentina's sizeable uranium supplies and its desire to decrease dependence on foreign suppliers of enriched uranium induced it to develop heavy-water natural uranium reactors. A different array of political and resource considerations persuaded Brazil and several other nations to pursue light-water technology. The technical characteristics of both reactor types pose a challenge to nonproliferation efforts. Both produce plutonium; and the light-water reactor cycle requires the use of enriched uranium as a fuel. However, the greater challenge arises from the political factors that will ultimately decide how these nuclear technologies are exploited; political intentions can be transformed from peaceful to militant by expanded technical capabilities.

Whether any Latin American country will turn its nuclear power industry into a source of armaments is impossible to predict with certainty. The region's principal effort to regulate its nuclear future – The Treaty of Tlatelolco – is probably the best guide to whether the Latin nations would accept nonproliferation measure such as a regional spent fuel storage center.

The 1967 Treaty of Tlatelolco establishes a regional nuclear-weapons-free zone and prohibits the production, possession, or foreign deployment of nuclear weapons within the 22 signatory countries. In addition, all nuclear-weapons signatories – France, the Netherlands, The U.K., and the U.S. – have signed protocols endorsing the Treaty's objectives.(22) Argentina, Brazil, and Chile are the only Latin nations that have refused to accept the Treaty without reservations.(23) Argentina and Brazil have declared their interest in employing peaceful nuclear explosives for various "development" purposes. Chile refuses to restrict its future nuclear options to promote nonproliferation, a goal that it believes serves the interests of only the developed, industrial nuclear-power nations.

A regional fuel cycle facility, starting with spent fuel storage objectives, could serve the Treaty of Tlatelolco by discrediting claims that indigenous enrichment and reprocessing are needed to avoid reliance on foreign sources of nuclear fuel. It could, over the longer term, be the world's first enrichment plant operated by nations that are not major exporters of nuclear equipment. This novel arrangement would obviously increase the chance that these nuclear consumer nations would become nuclear fuel exporters; they would be reprocessing more usable fuel than was needed domestically. However, export of enriched fuel would not have to occur if all members of the regional facility agreed to use the plant solely for regional nuclear programs. Even if they did not reach this agreement, approximately 15 years would pass before these nations could build a separation plant and become net exporters of enriched uranium. A Latin American regional storage facility, with reprocessing added in the 1990s and enrichment after the year 2000, need not hinder nonproliferation.

The chief obstacle to creating a regional storage facility that would be managed by the administrative agency of the Treaty of Tlatelolco (OPANAL) is that Argentina and Brazil – the two nations most able to develop nuclear weapons – remain outside the Treaty. A nuclear free zone in whose midst two states build nuclear weapons would be a mockery.

Argentina and Brazil have not announced plans to develop nuclear exports, but they clearly will have the potential to do so. Argentina is not likely to invest in a uranium enrichment plant because its natural uranium heavy-water reactors do not require enriched uranium. However, Brazil's light-water reactors do need enriched uranium. Brazil, therefore, might possibly build an enrichment facility and export excess nuclear fuel. Although West Germany has agreed to sell enrichment machinery to Brazil, international concern about proliferation may permanently block the actual transfer. If Brazil does not receive the West German equipment, it may prefer regionally-produced enriched uranium to continued reliance on foreign reactor fuel.

Should Argentina and Brazil opt for nuclear weapons, they would obviously want their own uranium enrichment and plutonium reprocessing facilities; should they not, they might well find regional storage for spent fuel advantageous. If OPANAL were to build an enrichment and storage plant, it would not only serve the energy independence and regional security desires of its members, but would clarify Argentine and Brazilian nuclear intentions by offering them a receptacle for spent fuel that would be inadequate only if they wished to develop nuclear weapons.

TOWARD FINDING AN ACCEPTABLE SITE

Selecting a specific site for fuel storage and/or reprocessing and waste disposal hinges on the question of political acceptability: first, it must be accepted by the rulers and population of the host country; and, equally important, it must be accepted by the governments of those countries which would be likely to use the facility.(24) Such acceptance, in turn, will hinge on whether the facility's system of controls and safeguards is beyond reproach, and whether the facility can provide an outlet for spent fuel that is less costly than domestic disposal.(25)

Some argue that a regional fuel cycle plant should be located on "internationalized" territory governed by a new international body with strong oversight powers. This approach, although attractive, would require such protracted international negotiations that it might deter prospective host countries from offering sites for the facility. Perhaps it would be easier if an existing international institution with some dispute-settlement powers administered the site.(26) Whatever the administrative approach selected, inspectors must be quaranteed full access to the plant.

Planners must also decide whether to locate the facility in one of the larger Latin American nations with an active nuclear power program or in a minor state that is unlikely to become a major discharger of spent fuel.

Some might argue that Mexico offers the ideal site, in view of the country's long-standing commitment to keeping Latin America totally free from nuclear weapons. Although Mexico offers sufficient fuel security, its ties with the U.S. to the point of supporting American views on nonproliferation(27), are too close for Latin American acceptability. Mexico was the architect of the Treaty of Tlatelolco and a strong supporter of the NPT; on the other hand, Argentina, Brazil, and Chile have risked conflict with the U.S. to retain the option to acquire nuclear explosive capabilities. Mexico's nuclear program depends largely on American equipment and fuel supplies, while Brazil and Argentina are committed to handling the entire fuel cycle domestically despite strong American objections. Chile's nuclear program is not yet sufficiently advanced to elicit American disapproval, but it, too, seems to emphasize national autonomy.(28)

Mexico's geographic location also makes it a poor site for a Latin American storage and reprocessing facility. Its distance from Latin American countries that would have significant amounts of spent fuel would increase transportation costs. In addition, Mexico is so close to the United States that it might want to join a North American multinational fuel cycle regime, with the plant located in Mexican territory.

Finding a site in any South American country now entering or likely to enter the era of nuclear power will not be easy; deep-seated political rivalries divide the continent.

Chile and Brazil, for example, are likely to veto Argentina as a host for a spent fuel management facility; Argentina's and Chile's overlapping territorial claims, particularly to certain islands located in the Beagle Channel, led to mobilization and redeployment of the two countries' armed forces in 1978. Since then tension appears to have diminished, but the fundamental dispute has not been resolved.

Argentina and Brazil not only dispute boundaries, but also compete for influence elsewhere in South America. This competition has become evident in disagreements about the construction of hydroelectric power plants in association with Paraguay on the Parana River, and in attempts by both countries to secure access to Bolivia's oil, natural gas, and iron ore.(29) The President of Argentina's Atomic Energy Commission has acknowledged that "there is indirect competition with Brazil" for nuclear technology in South America. This desire for primacy is demonstrated by Argentina's sale of reactors and other equipment to Peru and, possibly to Bolivia.(30)

Siting a nuclear fuel reprocessing facility in Brazil would evoke objections not only from Argentina but from Venezuela and Peru as well. Peru is hostile towards Brazilian road construction, and the associated future civilian and military activities, in the previously-uninhabited section of their common border. Venezuela also has unsettled border disputes with Brazil; in addition, it supported Guiana's territorial claims against Brazil and opposed the 1977 German-Brazilian nuclear agreement although it modified its position later that year.(31)

A Chilean site is not likely to find acceptance in Argentina, Bolivia, or Peru. There are major territorial claims outstanding among Peru, Bolivia, and Chile which can be traced back to the War of the Pacific a century ago. Bolivia recently broke diplomatic relations with Chile once again. Peru, the major loser in the War of the Pacific, has not relinquished its claims on territory it lost then or its suspicions of Chilean expansionism.

Siting in Peru would not only be rejected by Chile, but also by Colombia and Ecuador, again because of conflicting territorial claims. Neither country has given up its claim to land Peru captured in a 1940 border dispute.

Venezuela and Colombia would reject a site in each other's country because of disputes over territory and the Venezuelan treatment of illegal Colombian immigrants. Brazil might also find Venezuela unacceptable because of Venezuelan ambivalence on the German-Brazilian nuclear agreement and its support for American nonproliferation policy. Finally, Peru is likely to reject Ecuador, again because of continuing problems with border disputes. In addition, all the countries on South America's west cost are earthquake-prone.

The only South American countries remaining are Paraguay and Uruguay, two small buffer states, both parties to the Treaty of Tlatelolco and the NPT, and both promoters of regional nonproliferation.(32) These countries are the best candidate for hosting a regional nuclear fuel storage and reprocessing facility.

Paraguay has not indicated any interest in developing nuclear power; it will have adequate reserves of electricity from huge hydroelectric projects currently being constructed along the rivers that separate it from Brazil and Argentina. Paraguay owns one-half of the assets of these projects and will be entitled to one-half of the power generated. Paraguay appears to have no immediate use for the power generated in the 10,700 MW Itaipu hydroelectric complex currently under construction on the Paraguay-Brazil frontier; therefore, it plans to sell its quota to Brazil to repay the latter for financing construction costs. However, Paraguay can keep its share of the power when it is needed. Similar arrangements have been made with Argentina on the Corpus and Yacireta-Apipe hydroelectric generating stations. Given Paraguay's limited population and economic growth, the hydroelectric plants should provide adequate power reserves for the forseeable future.(33)

Paraguayan leader Alfredo Stroessner is probably the longest-serving president in Latin America, recently having won his sixth reelection. He has strong support from the armed forces, which he has molded into a tool of his interests, and from major factions of the dominant Colorado party. Paraguayan politics have traditionally been violent, and its leaders have demonstrated little concern for the basic political rights of the people. The Stroessner regime has, if anything, intensified this characteristic. Coercion, effectively combined with some socioeconomic development, and the government's politically-astute allocation of positions and resources have produced a long-term stability that is rare in Latin America. Although Paraguay's government is neither likeable nor commendable, its stability cannot be questioned.

Uruguay presents a somewhat different picture, both in nuclear development and political stability. Although it has a joint project with Argentina to build the Salto hydroelectric power plant, its lack of oil, coal, and natural gas, plus its higher level of development, would appear to make it a likely candidate for nuclear power plants in the not-too-distant future.

Like Paraguay, Uruguay is governed by its military establishment. Until the late 1960s, it was a stable, representative democracy that would have been an excellent host for a spent fuel management operation. In the late 1960s, however, Uruguay's historic stability was shattered by Tupamaros, an urban guerilla organization. Traditional counterinsurgency methods – massive detention, torture, executions without trial, and kidnappings – succeeded in wiping out the guerillas by the early 1970s, but they also sapped Uruguayan democracy of its vitality and strengthened military control. Uruguay's civilian president is unlikely to succeed himself; and the senior officer corps constitutes the real decision-making body.

Under the new regime the income distribution gap has widened and some welfare state features have been dismantled, but the GNP and the balance of payments have improved. Because it rests on a new political and economic foundation, Uruguay's future social tranquility is unpredictable.

There can be no guarantee that guerilla activity will not recur in Uruguay or that an effective subversive organization will not arise in Paraguay; nor can it be ensured that guerillas will not cause trouble in the U.S. or the Soviet Union. Both Paraguay and Uruguay have governments and internal security forces equal to the task of hosting a nuclear fuel storage facility. If one ignores the ethics of the methods used to achieve stability, Paraguay and Uruguay clearly offer the most secure political environments in Latin America for a spent fuel facility.

It is extremely difficult to ascertain whether Paraguay or Uruguay would want to host a spent fuel management facility. Uruguay has a research reactor and might decide to resort to nuclear-generated electricity. Its traditional hospitality to supranational and international agencies should help to generate support for this novel form of regional cooperation. Uruguayans are relatively well-educated, and some might object to a spent fuel facility on environmental grounds. However, these objections would not be decisive if the high-ranking officers and technocrats who are in control of the government wanted the plant.

Paraguay is not planning a nuclear program. The population's low level of education, and its traditional isolation from other countries diminish the chance of opposition from environmentalists. Environmental objections were raised in the Paraguayan legislature over the Itaipu hydroelectric projects;(34) but if the small ruling elite led by President Stroessner wanted the facility, lower-level overt opposition would evaporate. On the whole, Paraguay has been ignored by international agencies. Perhaps General Stroessner, after 24 years in power, might want to join the world-wide trend toward increasing international contacts. If so, locating a South American spent fuel management facility in Paraguay would be a high-visible accomplishment.

A major advantage of Paraguay and Uruguay as spent fuel facility sites is their regional acceptability. Neither country has serious boundary disputes or rivalries with potential participants. Nor can their territories be realistically considered as candidates for aggression from expansion-minded neighbors. Whatever international instability can be forecast in South America, Paraguay and Uruguay are not likely to be directly involved. Although liberal democracies may find Paraguay's and Uruguay's regimes unsavory, it should be remembered that representative democracy is uncommon to many of the Latin American nations potentially involved in international spent fuel storage. Finally, all major Western suppliers of nuclear equipment have normal diplomatic relations with both countries.

In a huge area such as Latin America, there is no universally-convenient location for a spent fuel facility. Long distances and natural obstacles – the Andes, for example – would mandate some circuitous spent fuel transport routes. Paraguay and Uruguay do not solve the transport problem, but they are conveniently located near Argentina and Brazil, the two nations moving most decisively toward the construction of nuclear power plants and the development of nuclear technology. If Mexico were to participate in a Latin regional scheme, transportation of its spent fuel would be burdensome. Instead, Mexico might join a North American fuel management facility, or host a multinational facility that could encompass Central America and the Caribbean; if that were to be done the overall transport problems of a South American regime would be greatly simplified.

Uruguay clearly has better access to countries with spent fuel than Paraguay. It possesses a long Atlantic coast, a major port, and a reasonably good highway and railway network; furthermore, there are very few natural barriers in the country. Spent fuel originating in countries with direct access to the Atlantic Ocean could be shipped directly to the port of Montevideo. Spent fuel from the Latin countries that border only on the Pacific might have to be shipped to either Valparaiso or Antofagasta in Chile, and from there by railroad or truck through Argentina into Uruguay.

Paraguay, lacking direct access to the sea, would present a more challenging transport problem. Countries with direct access to the Atlantic could transport spent fuel by ship through the Plata, Parana, and Paraguay Rivers into Paraguay's capital city of Asuncion or the port city of Encarnacion. This route, however, is slow and sometimes unpassable. It would be simpler to unload the spent fuel in Argentine or Brazilian ports and transport it from there to the fuel facility by railroad or truck. Shipments arriving from the Pacific could follow the same route as those destined for Uruguay, except that Antofagasta might be a more desirable transfer point than Valparaiso. Final selection of routes and transport methods would require exploring the cost, safety, and quality of existing systems, as well as the preferences of the host country.

Both Paraguay and Uruguay have the necessary transport infrastructure to host a multinational spent fuel storage facility. Both countries also have large areas that are sparsely settled, northern Uruguay and Paraguay's "Chaco" region, which makes them ideally suited to the needs of a spent fuel management facility.

TOWARD AN ACCEPTABLE SCHEME

It will not be easy to establish a multinational nuclear fuel cycle facility in Latin America. A host country must be found and mutually-acceptable control mechanisms must be devised – no small feat! Argentina, Brazil, and Chile have refused to renounce nuclear weapons and desire nuclear self-sufficiency. The facility would have to offer financial incentives to attract widespread participation from these and other countries.(35)

Selecting an existing regional organization to administer the site presents several advantages; these include cost, political acceptability, and the authority to settle disputes. Several regional organizations are potentially suited for the task. The Organization of American States, through its Inter-American Nuclear Energy Commission, could administer a spent fuel facility; but its close ties with the United States would probably make the OAS unacceptable to those potential participants, such as Brazil and Argentina, that have collided with the United States on nuclear policy.(36) The Latin American Economic Organization does not include the United States, but it does include Cuba, an unlikely participant; it also includes several small nations whose interest would be marginal.

The Treaty of Tlatelolco's administrative agency, the Organization for the Prohibition of Nuclear Weapons in Latin America (OPANAL), could potentially manage the regional spent fuel facility. OPANAL was created in 1969 as the administrative body of the Treaty; situated in Mexico City, it is comprised of three institutions, the General Conference, the Council, and the Secretariat. The General Conference has representatives from the 22 signatories; it allocates the budget, determines protocol for implementing Treaty guidelines, and sets major policy. The Council conducts special "demand" inspections in accordance with Article 10 of the Treaty; and the Secretariat handles other duties.

Most nations that are party to the Treaty also signed the worldwide Non-Proliferation Treaty (NPT), which went into effect in 1970. Most also have bilateral arrangements with the IAEA permitting inspection of their nuclear facilities. Despite the fact that it was established 10 years ago, OPANAL is capable of accomodating the growing nuclear needs of the region. Colombia, Mexico, Peru, and Venezuela – all parties to Tlatelolco – are expanding their nuclear power programs. If the nuclear needs of the region continue to grow and consensus is developed to support the creation of a regional fuel cycle facility, OPANAL could logically serve as its administrative arm.(37)

A successful OPANAL spent fuel program, however, would require Argentina, Chile, and Brazil to become parties to the Treaty. Until that happens, entrusting OPANAL with the management of the regional plant would exclude the region's leading nuclear states. One cannot object to the suggestion that the Organization "should be carefully cultivated and encouraged as an important instrument in the pursuit of world order."(38) Nevertheless, its current membership disqualifies it from administering a regional spent fuel facility.

A better vehicle might be the Latin American Free Trade Association, whose administrative structure is headquartered in Montevideo. Its members are Argentina, Bolivia, Brazil, Chile, Colombia, Ecuador, Mexico, Paraguay, Uruguay, and Venezuela – all the Latin American countries likely to need spent fuel disposal in the forseeable future, as well the two countries best suited to hosting the facility. LAFTA, originally founded to lower regional trade barriers, has had no major successes since 1967; it could provide an administrative umbrella and a complete membership list, thus settling <u>de facto</u> the problem of membership and avoiding problems related to the juridical status of a new supranational entity.

The direct responsibility for protecting the site would fall on the military establishment of the host country, perhaps on a specific unit permanently assigned to this mission. Although the armed forces of Paraguay and Uruguay are small, their respective sizes appear to exceed their potential external threats. Their roles over the last 80 years have been limited to coups d'etat and domestic counterinsurgency; however, in view of the present political role of these military establishments, future coups d'etat would seem to be unlikely. Thus the military establishments of the potential host countries might welcome the chance for a new mission. The facility's security unit would be specially equipped with modern material and would receive financial support from all member countries. This support could be a real attraction to countries such as Paraguay and Uruguay where approximately $1500 and $2800, respectively, is spent per year for each member of the armed forces.

Each country storing spent fuel in the proposed facility should have the right to appoint a small number of inspectors with the authority to supervise all aspects of the plant. This right might prove attractive to countries interested in following their neighbors' nuclear activities. It is even possible that some countries not originally inclined to participate in regional spent fuel storage might join to gain this "right of oversight." Needless to say, the national inspectors should also be able to verify the storage of their country's spent fuel and prevent its improper diversion. Finally, each participating country must have guarantees that its spent fuel will be returned on demand, and the financial responsibility for return transportation must be clearly outlined.

It seems that the time has come to seriously consider the establishment of at least one regional spent fuel storage facility in Latin America. Informal contacts with policy-makers of the affected countries, as well as the general tenor of their nuclear programs, leave the impression that they may be receptive to the concept of a multinational facility located in the region. The specific details of such an entity would, no doubt, be subject to the give and take of international negotiations. The final outcome would also be based on the intentions of the Latin American countries with access to nuclear technology, and on the incentives provided by nations pursuing nonproliferation policies. The present preliminary framework does not hinder the legitimate technological goals of potential participants, nor will their political autonomy be mangled by the necessity to subscribe to public renunciation of their sovereign right to manufacture any type of weapons. In fact, the ruling elite of most Latin American countries are likely to find a great deal of merit in regional nonproliferation; however, political considerations may prevent them from acknowledging it. A regionally-operated and controlled spent fuel storage facility may very well provide a de facto mechanism that would discourage the production of nuclear weapons without public promises or technological self-denials.

NOTES

(1) La Nacion (Buenos Aires), January 18, 1978, p. 7. The latest contract, whose approval by the Argentine government was reported in the item, provided for the supply of 715 elements, with an option to purchase an additional 195.

(2) General Osiris G. Villegas, "Puntos de Vista para una Politica Nuclear Nacional," Estrategia, July-December, 1976, p. 11. Also see La Opinion (Buenos Aires), January 14, 1978, p. 9; this newspaper, under direct government supervision, refers to a recent study of Argentina's Atomic Energy Commission which contemplates the shift to German reactors. It also raises the possibility of asking for "most favored nation" treatment in order to by-pass the Suppliers' Group Agreement. Comments made by the President of the National Commission of Atomic Energy in August 1978, indicated that the problem had not been solved; see Jornal do Brasil, August 27, 1978, special section, p. 1.

(3) As reported in Jorge A. Sabato and Raul J. Frydman, "La Energia Nuclear en American Latina," in Jorge A. Sabato and others, Energia Nuclear, una Opcion para el Desarrollo (Buenos Aires: Tierra Nueva, 1977), p. 90. One-third of these reserves are proven.

(4) Osiris Villegas writes that the decision to build the heavy water pilot plant was made in April 1976. See his op. cit., p. 11. According to John R. Redick, intentions to build the plant had been reported earlier; see his "Nuclear Proliferation in Latin America," in Roger W. Fontaine and James D. Theberge (eds.), Latin America's New Internationalism: The End of Hemispheric Isolation (New York: Praeger, 1976), p. 274. Newspaper reports indicate that the design was almost completed by the end of 1977; see La Opinion (Buenos Aires), December 29, 1977, p. 12, and La Nacion (Buenos Aires), December 27, 1977, section 2, p. 1. The amount of plutonium has been calculated by the authors on the basis of AIEA figures.

(5) Possession of the "technological basis" was claimed by Argentine diplomats during the debate that preceded the approval of the NPT; for a report by Argentina's chief negotiator see Jose Maria Ruda, "La Posicion Argentina en Cuanto al Tratado sobre la No Proliferacion de las Armas Nucleares," Estrategia, September 1970-February 1971, pp. 75-80. After Secretary of State Vance's visit in November 1977, the Argentines announced that they had the capacity to produce nuclear weapons, but were not interested in doing so, according to Latin America Political Report, November 25, 1977, 11: 364-65.

(6) Argentina has announced more than once its readiness to ratify the Treaty of Tlatelolco; the latest announcement was included in the joint communique issued at the end of the November 1977 visit of Secretary of State Cyrus Vance. See Argentina (Washington, D.C.), January 1978, p. 1; and Latin America Political Report, November 25, 1977, 11: 364-65.

For an extended discussion of Argentine options, see C.H. Waisman, "Incentives for Nuclear Proliferation: The Case of Argentina," in Onkar Marwah and Ann Schulz (eds.), Nuclear Proliferation and the Near-Nuclear Countries (Cambridge, Mass.: Ballinger, 1975), pp. 279-293; and Astiz, "The Military and Brazil's Geopolitical Mission," paper delivered at the Symposium on Recent Developments in the Political Role of the South American Military, The University of New Mexico, April 4-6, 1974.

(7) General Osiris Villegas, op. cit., p. 12 (emphasis added).

(8) Reported in La Nacion (Buenos Aires), April 13, 1978, p. 3.

(9) Roger Fontaine and James Theberge, Latin America's New Internationalism, p. 278. Brazilian foreign policy analysts raised the issue of "nuclear dependency" as far back as the 1960s. See, for instance, Jose Honorio Rodrigues, Interesse Nacional e Politica Externa (Rio de Janeiro: Civilizacao Brasileira, 1966) p. 213.

(10) Nature, "Brazil Plans 63 Nuclear Reactors This Century," Vol. 257, October 9, 1975, p. 437. For a less optimistic view, see The Washington Post, October 15, 1978, p. A-27.

(11) Nuclear Engineering International, April 1978, p. 3. The confusing relationship between Brazil and Urenco has continued; see, for example, Jornal do Brasil (Rio de Janeiro), August 17, 1978, section 1, p. 27.

(12) The first reactor is mentioned by Redick, "Nuclear Proliferation...," p. 287. The second reactor was reported in Que Pasa (Santiago), October 27-November 2, 1977, p. 11, and in Latin America Political Report, October 21, 1977, 11: 328.

(13) Redick, op. cit., p. 287; "U.S. Nuclear Policy and Latin America" (Charlottesville, Va.: Vantage Conference Report, Dec. 1012, 1976), p. 22.

(14) Redick "Nuclear Proliferation. . .," pp. 289-90; Sabato and others, op. cit., p. 97.

(15) Nuclear News Buyers Guide, Mid-February 1978, p. 54.

(16) James D. Theberge and Roger W. Fontaine, Latin America: Struggle for Progress, volume XIV of Critical Choices for Americans (Lexington, Mass.: Lexington Books, 1977), p. 23. For a description of Peru's hydroelectric potential, see Caretas (Lima) September 18, 1978, pp. 64-66.

(17) Latin American Political Report, March 11, 1977, 11: 77. A previous agreement of mutual cooperation in nuclear energy, signed in May 1968, did not lead to meaningful cooperation.

(18) La Opinion (Buenos Aires), January 4, 1978, p. 10.

(19) The economic difficulties of the Peruvian military regime have been widely publicized; see The New York Times, December 17, 1977, p. 1, and December 18, 1977, p. F-15.

(20) Redick, "Nuclear Proliferation...," p. 288.

(21) "U.S. Nuclear Policy...," p. 23.

(22) "Additional protocols to the Treaty concern the respect and adherence by non-Latin nations with territorial interests in the Americas and nuclear weapons states." "U.S. Nuclear Policy and Latin America," Vantage Conference Report, p. 24. Great Britain, the U.S. and France are affected by the first protocol. Protocol 2 prohibiting the use of nuclear weapons against nations party to Tlatelolco was signed by the People's Republic of China, the U.S., France, and Great Britain.

(23) All OAS members are also members of IANEC.

(24) Detailed consideration of the siting question is presented by David Deese and Frederick B. William in Chapter 1 in this volume.

(25) The economic aspects of international spent fuel storage are analyzed by Boyce Greer and Mark Dalzell in Chapter 8 in this volume.

(26) Legal and administrative questions are discussed in Deese and Williams.

(27) For a more comprehensive view of Mexico's foreign policy, see Astiz, "Mexico's Foreign Policy: Disguised Dependency," Current History, 66: 220-225, May 1974.

(28) Redick has an interesting discussion on this matter; see his "Regional Nuclear Arms Control in Latin America," International Organization, 29: 415-445, Spring 1975. However, the claim made on p. 429 that Chile ratified the Treaty of Tlatelolco in 1974 because of Mexican pressure appears to be questionable, considering the disagreements between the Chilean military regime that overthrew Allende and the Echeverria administration.

(29) For a discussion of Argentine-Brazilian rivalries, see Gregory F. Treverton, "Latin America in World Politics: The Next Decade," Adelphi Paper No. 137 (London: IISS, 1977), pp. 29-45; and Astiz, "The Military...," passim.

(30) La Opinion (Buenos Aires), December 29, 1977, p. 12. Transfers of

nuclear technology from Argentina to Peru and Bolivia are mentioned earlier in this chapter.

(31) Latin America Political Report, April 1, 1977, 11: 98-100, and October 28, 1977, 11: 330-332. Consult also Alexandre S.C. Barros, "The Diplomacy of National Security: South American International Relations in a Defrosting World," in Ronald G. Hellman and H. Jon Rosenbaum (eds.), Latin America: The Search for a New International Role (New York: Sage Publications, 1975), pp. 131-150.

(32) Unknown to the authors at the time the first draft of this paper was written, Paraguay has already been suggested as the site of a regional spent fuel facility. See "U.S. Nuclear Policy...," pp. 30-31.

(33) The chief executive officer of Paraguay's power authority (ANDE) stated unequivocally that it was not economically feasible to resort to nuclear energy in his country. See his presentation before the Paraguayan Senate, reproduced in Efrain Enriquez Gamon, Itaipu; Aguas que Valen Oro (Buenos Aires: Grafica Guadalupe, 1975), pp. 196-224; his remarks on nuclear power appear on p. 212.

(34) See the parliamentary debates in Gamon, op. cit., pp. 79-447.

(35) See Greer and Dalzell.

(36) When the Treaty of Tlatelolco materialized, it was kept clearly separated from the OAS and its Inter-American Nuclear Energy Commission. This attitude does not appear to have changed. See Redick, "Regional Nuclear Arms Control...," pp. 442-43.

(37) Redick, "Regional Nuclear Arms Control. . .," p. 444.

(38) The suggestion has been made, somewhat vaguely, in "U.S. Nuclear Policy. . .," p. 31.

5 Asia and the Middle East

Richard Broinowski

With the exception of Japan, the Asian and Middle Eastern countries embarking upon nuclear power programs have devoted little constructive thought to the problems of long-term storage of spent nuclear fuel. On the contrary, problems relating to the back end of the nuclear fuel cycle have been shunted aside in the rush to develop nuclear power as a partial substitute for oil-fired generators. Most Third World countries involved in nuclear programs want to develop this energy source as quickly as possible to reduce growing oil import bills, assure independence of energy supplies, and take part in advanced scientific development.

This chapter surveys the attitudes among Asian, Pacific Basin, and Middle Eastern countries toward the problem of spent fuel management. In particular, it examines the feasibility of developing cooperative regional spent fuel storage regimes. Some of the information used herein comes from available publications, but most was obtained from scientific attaches and senior political officers representing their countries in the Washington embassies. In general, the diplomatic representatives interviewed were negotiating regularly with United States agencies over their countries' fledgling nuclear-power programs, or were involved with the International Nuclear Fuel Cycle Evaluation (INFCE).

Countries covered in this survey include Japan, Republic of Korea, and Taiwan in East Asia; Thailand, Malaysia, Singapore, the Philippines, and Indonesia in Southeast Asia; Australia; and Egypt, Iran, Iraq, Pakistan, Israel and Libya in the Middle East. These are by no means all the countries in Asia and the Middle East now contemplating nuclear power. According to a 1977 United States Office of Technology Assessment report, Kuwait, Morocco, Tunisia, and Saudi Arabia in the Middle East, and Hong Kong in East Asia may also build reactors before the end of the century. Discussing the spent fuel storage options available to these countries before they commit themselves to nuclear power would be pointless; therefore, they have not been included in this survey.

The International Atomic Energy Agency has played a major role in raising the awareness of Third World countries to nuclear power's potential. IAEA has surveyed most of the Asian and Middle Eastern countries discussed in this chapter to establish what the Agency considers their future nuclear power requirements. These surveys have sometimes exaggerated the projected role of nuclear power, particularly in the Republic of Korea, the Philippines, Indonesia, and Pakistan. In Pakistan, for example, a 1975 IAEA survey recommended that the country should construct 24 600 MW reactors by the year 2000. However, economic realities have to be faced: the cost of constructing nuclear reactors is escalating; heavy balance-of-payments problems are caused by buying such large items abroad; it is difficult to know the eventual costs of the storage, transportation, and disposal of spent nuclear fuel; and reactors must eventually be decommissioned. In view of Pakistan's relatively small future energy requirements, this IAEA recommendation was probably inflated. Nevertheless, IAEA-engendered enthusiasm has infected many Third World government energy diversification programs, and some countries are now reluctantly scaling down their reactor construction programs in the face of prohibitive costs.

Several countries are taking a hard second look at optimistic projections for nuclear power. One of these is Singapore; the island republic, with IAEA assistance, carried out a survey on the possibility of constructing at least one reactor, but it decided that no suitably safe site exists on the crowded island. The government is still considering a floating reactor. Malaysia has calculated that the huge initial capital cost of reactor construction outweighs any possible long-term advantage of having an alternative source of power when Malaysian oil runs out. In what may become a key example of LDC concern about environmental pollution, Thailand has shelved plans for installing reactors, at least in the near future, because thermal pollution hurts fishing in the shallow waters of the Gulf of Siam.

A fortunate few of the developing countries surveyed do not need to forsake petroleum dependence in the near future. Iran, Indonesia, Libya, Iraq, and Algeria are planning nuclear energy programs as much to boost national prestige as to ensure continued energy supplies once their petroleum runs out. If the oil states with the highest per capita oil incomes – Saudi Arabia, Kuwait, and the United Arab Emirates – ever decide to invest in nuclear power programs, their motives will relate more to divesting themselves of surplus petrodollars than any need to diversify energy supplies.

For most of the countries this chapter surveys, however, long-term plans envisage that nuclear power will provide between 10 and 40 percent of their total electricity by the year 2000. They will have accumulated spent nuclear fuel, and they will have to determine what to do with it.

Before looking at individual countries, here are some general observations about nuclear waste management in Asia and the Mideast:

First, among most countries surveyed, interest in the deliberations of the International Nuclear Fuel Cycle Evaluation remains high. The unanimous view of Washington-based diplomats is that the Evaluation has provided a breathing space during which governments can delay making hard policy decisions about the future of their nuclear power programs until they find out more about other countries' plans.

Second, there is little enthusiasm for releasing spent nuclear fuel into an international storage area unless the donor countries can retrieve the fuel for reprocessing upon request.

Third, if international storage becomes necessary or desirable, most countries would prefer regional arrangements to storage in the United States.

Fourth, no country surveyed has coherent plans for national storage of spent fuel beyond the 1990 to 2000 range. Most governments have planned no further than the temporary storage capacities being built into their reactors by commercial contractors. They have speculated little about what lies beyond. No country surveyed, except Japan, has figured out the cost of long-term spent fuel storage. Feasibility studies by commercial contractors justifying nuclear installations tend to omit such open-ended figures from their cost-benefit calculations.

Fifth, although the representatives of several countries claim that environmental opposition to nuclear energy has peaked in their countries, only Pakistan and the Philippines suggest that regional or international spent fuel storage facilities might be established within their own national borders. Governments have either not considered the possibility, or are convinced that hosting such a regime would arouse heated opposition from environmentalists.

Sixth, there is a reasonably high chance of establishing multilateral oversight of national storage facilities, but each arrangement must be examined individually.

EAST ASIA

China

One country completely close-mouthed about its nuclear-power plans is the People's Republic of China. It appears reasonably certain, however, that China has so far confined its nuclear program to military applications. The Chinese nuclear weapons program is now two decades old. In its first decade, the basic scientific and production facilities were established. The Chinese, with Soviet aid, built the Lop Nor test area and the Lanchon gaseous diffusion plants. After the Sino-Soviet rift at the start of the 1960s, China continued to expand its atomic weapons facilities: The Lanchon plant has been producing weapons grade U235 since 1963, and plutonium production began at the Yumen reactor complex in 1967; a second gaseous diffusion plant may have been established, and perhaps additional plutonium facilities as well.

Although no definite information about civilian nuclear power systems has come to the West, China has shown interest in various power reactors and will eventually be able to apply its military technology to civilian applications. China is not a party to the NPT, and it is extremely unlikely that China would be interested in contributing to a regional or international spent fuel storage regime once it starts its own civil power program. Nor is it known at this stage what China's attitude might be concerning participation in such a regime by other states with nuclear power facilities.

Japan

Because of its status as an economic superpower and its already extensive nuclear reactor network, Japan's attitude toward regional and international storage regimes may strongly influence the positions of less developed nuclear countries, especially in East and Southeast Asia.

Japan does not now appear willing to contribute to an international or regional storage regime. Japan's policy is to consolidate and expand the reprocessing technology it has developed; it plans to manufacture plutonium for its breeder program and to curtail uranium imports by "stretching" existing uranium fuel for LWRs. The Tokai Mura reprocessing plant will become fully operational in September 1979 after two years of experimental running. Even at full production, however, the plant will be able to reprocess only 2,300 metric tons of the 7,400 metric tons of spent fuel that Japan estimates it will generate by 1990. Rather than consigning the remainder to some future spent fuel storage regime, Japan will probably rely on toll contracts with European companies to reprocess the residue. A second Japanese reprocessing facility may be constructed to come on stream in 1990 to handle future waste.

Thus, Japan would not favor contributing its spent fuel to a regional international storage regime. Indeed, as a matter of its own regional standing, Japan could hardly store spent fuel in other countries without being accused of exporting its polluting industries. Nor, it seems, would the Japanese favor allowing any part of Japan to be used as a storage site for the spent fuel of other countries; the political and environmental difficulties would be enormous. Furthermore, given Japan's high population density and susceptibility to earthquakes, finding a suitable Japanese site might be impossible.

On the other hand, Japan is well aware of the dangers of proliferation and might agree to support a system under which spent fuel from other countries could be stored regionally or internationally, and perhaps reprocessed. However, Japan will take no definite position until the INFCE deliberations are concluded. The Japanese strongly support INFCE and will play a positive role in its work. They will also continue to take an active part in international task forces attempting to find technical solutions to spent fuel and waste storage problems. The Japanese Power Reactor and Nuclear Fuel Development Corporation has examined the French solidification program and is now studying waste solidification techniques.

South Korea

Escalating costs have forced the ROK to shrink projections for its nuclear power program. The program nevertheless remains ambitious, as the ROK intends to have five major power plants operating by 1986: One Westinghouse PWR of 595 MW is presently being test-run; a Canadian CANDU of 600 MW, and another Westinghouse PWR of 600 MW will be on stream by 1983. South Korea has its own, albeit low grade, uranium deposits. Concern that the ROK would start its own reprocessing receded in 1975 when Seoul cancelled its contract to purchase a pilot reprocessing plant from France. The decision to cancel was influenced by American disapproval, and by negotiations then in process with Canada for financing the purchase of a CANDU reactor. South Korea has ratified the NPT and government spokesmen deny any intention to develop a nuclear weapons capability.

Although it has no concrete plans for long-term fuel storage, the Koreans might be willing to contribute toward an international or regional spent fuel storage regime, or they might agree to return their spent fuel to the U.S. They would, however, probably want to retain the option to retrieve their spent fuel for reprocessing.

Taiwan

Taiwan's capacity for generating nuclear energy is expanding rapidly. Its nuclear industry began with a small research reactor provided by the United States under a bilateral agreement. Six LWRs are planned in the Taiwanese program – two GE BWRs of 600 MW, two GE BWRs of 950 MW, and two Westinghouse PWRs of 900 MW. The first of these, a 600 MW GE reactor, became critical in November 1977; the other five are scheduled to start up in 1978, 1980, 1981, 1983, and 1984.

Perhaps Taiwan's policy best reveals the ability of the U.S., through quiet, deliberate, and persistent diplomacy, to persuade a country to abandon its apparent plans to build nuclear weapons. Taiwan had built an experimental reprocessing plant at its Institute of Nuclear Energy Research, which also had two research reactors. The reprocessing plant had been run as a "cold" plant prior to Carter Administration objections. In response to these objections, the reprocessing plant has reportedly been closed down and the laboratory dismantled.

Taiwan appears, therefore, to be walking in step with U.S. antiproliferation policies. According to the Taiwanese Scientific Attache in Washington, Taiwan will return spent nuclear fuel to the United States if requested, or to any international or regional location designated for that purpose. But Taiwan has held no formal discussion with the United States or other countries in the region on cooperative arrangements for spent fuel storage or waste disposal. Taiwan would probably favor a regional storage arrangement, although the government appears to be in no particular hurry and has designed its reactors to accommodate up to 10 years of spent fuel on site.

SOUTHEAST ASIA

The countries surveyed include the five members of ASEAN, the Philippines, Indonesia, Thailand, Malaysia, and Singapore. Their combined population is about 230 million people. With the exception of Singapore, a prosperous city-state, their per capita incomes are among the lowest in the Third World. However, their potential economic growth is enormous, as is their potential demand for electric power. Through regional cooperation, they could in time develop into a powerful economic bloc in which the prospects of sharing energy resources, including nuclear facilities, could be a significant feature. ASEAN already has tentative schemes to integrate some industries, share foodstuffs and petroleum in emergencies, and establish reciprocal trade concessions. In the very long term, this mutual assistance could lead to a level of cohesion and interdependence similar to the EEC's.

The Philippines is the only ASEAN country now committed to nuclear power, although Indonesia may shortly follow. To avoid purchasing enriched uranium, Indonesia is considering buying a CANDU reactor, which could be operative by 1985. An Italian company is doing preliminary feasibility studies. However, the Philippines, Indonesia, and Thailand all have research reactors; and Malaysia is presently negotiating for one from the U.S.

The other countries of Southeast Asia – Burma, Laos, Cambodia and Vietnam – appear to have no nuclear development plans. Vietnam, in view of its industrial base and technological development, would be the most likely to seek nuclear power, but ASEAN's Vietnam watchers agree that it will take at least five years of economic reconstruction before Vietnam can possibly embark on such expensive innovations.

The Philippines

The Philippines is the first of the ASEAN countries to construct a nuclear power reactor, a 620 MW Westinghouse PWR at the Bataan Industrial Zone on the western side of Manila Bay. Despite recent furors over alleged bribes and kickbacks paid by Westinghouse to the Filipino contractor, Disini, despite criticism of the reactor's safety features by the American Union of Concerned Scientists and some loose talk in Manila about handing the entire contract over to the Russians, it is safe to assume that Westinghouse will complete the reactor on schedule sometime in 1982. A site has already been excavated for a second reactor to go up next to the first, but the Marcos government has become very worried about costs: The price of the first reactor has risen from early estimates of $300 million to a current $1.2 billion.

In April 1978, Dr. Librado Ibe of the Philippine Atomic Energy Commission, and Dr. Higino Ibarra of the Philippine National Power Corporation's Nuclear Services Division said that the PAEC was actively considering hosting a regional spent fuel storage facility, and possibly a reprocessing plant. The PAEC has already examined 10 locations in the archipelago for their accessibility, security, and geological stability. The Bataan Industrial Zone, and the western island of Palawan appear to top the list. Official Philippine thinking indicates that such a regional regime would not be confined to the ASEAN countries, but could service Taiwan and South Korea as well. There is nothing in the ASEAN charter to prevent this; in fact, such an arrangement could conceivably serve as a precedent to expand ASEAN membership to other countries in the region. Drs. Ibe and Ibarra have said the idea would have to be exhaustively researched and that no decision would be reached until the conclusion of the current International Fuel Cycle Evaluation.

ASEAN councils have probably not devoted much discussion to a regional spent fuel storage regime. Nevertheless, the fact that ASEAN's member countries have a strong desire to put economic substance into their organization and the Philippine Government wants to remain an innovative leader both within the organization and in Asia generally will probably keep the idea alive. Perhaps it will be implemented as the region's nuclear power programs develop to the point where spent fuel storage facilities are filled. The environmental and safety aspects of such a proposal will be considered later in the chapter.

Indonesia

Indonesia's Atomic Energy Institute operates two small research reactors, one at Jogjakarta and the other near Bandung. The country has plans to develop a significant power program by the mid-1980s; two possible sites have been chosen, one at Mount Muria in north central Java and the other on the south coast of Java. Indonesian officials have talked about nuclear power cooperation with Malaysia and Singapore, so far without results. Although ASEAN officials have apparently not given serious thought to a regional spent fuel storage regime, the Indonesians would probably welcome cooperation in this area to strengthen the pact. Like the Filipinos, they have a strong commitment to ASEAN. However, Indonesia might not consent to a Philippine spent fuel facility. With its huge population, large land area, and abundant natural resources, Indonesia considers itself *primus inter pares* within ASEAN.

Thailand

Thailand commissioned a nuclear power feasibility study in 1973. The study recommended constructing a nuclear reactor, but the civilian Cabinet of Sisawant Pua-Wongpet rejected it for environmental, economic, security, and technical reasons; this was the first time that a Third World country had rejected a nuclear power program. A strong pro nuclear power lobby has been more vocal since a military government resumed in 1975, but the nuclear study languishes nonetheless.

Malaysia

Malaysia, sitting atop plenty of oil, has no present plans for a nuclear power program. However, the government is negotiating with the U.S. to purchase Malaysia's first experimental reactor, a TRIGA III.

Singapore

Singapore had tentative plans for a floating nuclear power reactor, but the idea has been shelved as too expensive and impractical. A city-state with a population density approaching that of Hong Kong, Singapore has virtually no place to put a reactor. Perhaps at some more advanced stage of ASEAN cooperation Indonesia might lease an island to Singapore for siting a reactor.

AUSTRALIA

The introduction of nuclear power into ASEAN is a recent phenomenon; only one or two reactors will be operational by the mid-1980s. However, ASEAN governments are already speculating about regional cooperation in at least three areas. The first is the distant possibility that Bataan or the island of Palawan in the Philippines might eventually host an ASEAN regional reprocessing and storage site. The second is that Indonesia might share power and perhaps storage arrangements with Malaysia and Singapore. The third is that the ASEAN countries may increasingly turn to Australia as a regional source of raw uranium, as an enrichment and fabricating base for uranium fuel, and as a site for the long-term storage of spent fuel and/or reprocessing. The Philippines has officially told Australia that it will depend upon Australian uranium for all its nuclear fuel requirements, except possibly the first charge for its reactor; Australian uranium may not be available to allow enrichment and fabrication before the scheduled start-up in 1982. Australian advances in nuclear technology could also serve regional interests at some future time.

The Australian Government is well aware of its potential role as a world supplier of uranium, a regional supplier of nuclear technology, and base for long-term spent fuel storage and/or reprocessing. It qualifies for the role on all three counts.

First, Australia has perhaps 20 percent of the Western World's proven low-cost uranium reserves. A 1976 U.S. Energy Research and Development Administration study estimated that Australia had 430,000 tons of "reasonably assured" uranium oxide and 104,000 tons of "estimated additional" reserves.

Second, although it shelved a plan to build a 500 MW reactor at Jervis Bay in New South Wales in 1969, and has no plans to resurrect the venture, Australia has a sophisticated nuclear research program at Lucas Heights in Sydney. Equipped with two research reactors, one an 11 MW and one a 10 MW, the facility also has a working pilot enrichment plant of its own cascade design capable of producing a kilogram of enriched uranium a day.

Third, with its vast land area, dryness, sparse population, and ancient and stable granite structures, Australia is perhaps one of the world's most uniquely-qualified countries to host a long-term international or regional nuclear spent fuel storage regime and/or a regional reprocessing and waste disposal area. Some Australian nuclear scientists and economists have argued that Australia should do so on commercial, as well as research and development, grounds.

Yet, because of domestic opposition, Australia has hesitated to develop its potential in any of these three areas. It was only after Malcolm Praser's Conservative government was returned to office in December 1977 that the government decided to mine the rich Alligator River uranium reserves in the Northern Territory. Mr. Fraser will, however, continue to face vehement domestic opposition; the last national poll showed 53 percent of Australians support this uranium export policy, but the number of those opposed to any uranium sales whatsoever has climed from 34 to 42 percent since June 1977.

The powerful Australian Council of Trade Unions (ACTU) continues to oppose opening new uranium mines until it is satisfed that international safeguards to prevent nuclear weapons proliferation, and spent fuel storage techniques are adequate.At a special conference in March 1978, the ACTU agreed to allow the government to honor existing uranium contracts – presumably from stockpiles held by the Australian Atomic Energy Commission and from Australia's only operating uranium mine at Mary Kathleen in Queensland. However it voted against the opening of new mines until safeguards and storage were ensured.

Under such conditions it is unlikely that the Australian Government will relax previously-announced stringent controls, reached through bilateral agreements, governing the disposal of Australian uranium, even when this leads to multi-labeling.

Continued strong domestic opposition to uranium exports will prevent consideration of the idea that spent Australian fuel should be returned to Australia after being leased rather than sold to customers for storage. In a statement to Parliament in Canberra following the Praser government's decision to export uranium, the Minister for Environment, Kevin Newman, made the following statement:

> The responsibility for disposing, in an environmentally-responsible manner, of waste arising from nuclear power generation in countries abroad, is a matter for those countries which generate electricity by nuclear means. There is no intention of Australia storing other countries' radioactive wastes.

Time alone will tell how strong this resolve will remain. Now that the government is seeking Parliamentary approval of legislation, and creating a Uranium Mining Authority to supervise the sale of uranium, strong political reasons require adherence to it.

THE MIDDLE EAST

Several countries in the Middle East have nuclear research reactors and research programs: Egypt has a Soviet-constructed two megawatt reactor at Inshass; Iraq has a Soviet research reactor – located south of Baghdad – that was constructed in 1968; Israel has two experimental reactors, one constructed by the U.S. in 1955 and the other by France in 1957; Libya has a 10 MW Soviet reactor.

Egypt, Libya, Iran, and – presumably – Israel have begun nuclear power programs or have definite commitments to do so. Algeria, Morocco, Iraq, Saudi Arabia, Kuwait, and – perhaps – the United Arab Emirates may eventually embark on nuclear power programs. Because of its close geopolitical ties to Iran and the countries of the Persian Gulf, and because it has committed itself to a nuclear power program, Pakistan should be included in a discussion of the potential for a Middle Eastern regional spent fuel storage/reprocessing regime.

Egypt

Of the five committed countries, Egypt's and Libya's programs best meet nonproliferation criteria. Egypt is an example of where American nonproliferation diplomacy has scored well. In 1976, Egypt signed an agreement with the U.S. for nuclear power plants capable of generating 2000 to 3000 megawatts; the Egyptians subsequently signed a contract with Westinghouse for two large PWR reactors, the first of which is scheduled to come on stream in 1983. Construction on the plants will begin as soon as the U.S. Congress has approved the agreement, which could prove to be a model for the Western world. Under it, the U.S. will take back all spent fuel from the reactors and store, dispose, or reprocess it at American discretion. This arrangement reassures Egypt's neighbors that it does not intend to use the reactors for developing atomic weapons. According to the Egyptian Embassy in Washington, a scientific committee in the Arab League may eventually develop some regional nuclear arrangement, both for storage and reprocessing; but this is far in the future. In the meantime, Egypt has agreed to support President Carter's proposal to establish an international fuel supply and retrieval system.

Libya

Libya is another country whose program will probably meet nonproliferation criteria. Although precise details are unavailable, it appears that the Soviet Union has agreed to construct a 440 MW reactor in Libya at a cost of some $330 million. There has been no official announcement about this plant, but there is little doubt that it will be placed under IAEA safeguards and that the USSR will follow the same practice it has adopted in Eastern Europe: All spent fuel will be returned to the Soviet Union.

Iran

Iran has a program to develop about 20 nuclear power stations by the end of the century. Although a drop in Iran's oil revenues has slowed the pace of reactor acquisition, the total program remains unchanged. The country has firm contracts for four LWR's: Two 1200 MW reactors are scheduled to come from West Germany in 1980 and 1981; and two French Framatome 900 MW reactors are due for commissioning in 1982 and 1983. In addition to negotiations with American companies that may result in the acquisition of eight Westinghouse PWRs, Iran is negotiating with Italian, Japanese, and Swedish companies to reduce its dependence on any one country for nuclear technology. Iran has considered a national reprocessing facility, but is now exploring bilateral and multilateral reprocessing arrangements. In February 1977, Iran and India signed a protocol on the peaceful uses of atomic energy. Although Iran is party to the NPT, the Shah has stated that Iran may be forced to acquire nuclear weapons if other states do so.

Iran is ambivalent about contributing to a regional or international spent fuel storage regime. The country "went along with" America's suggestions on the storage of spent nuclear fuel, but the Shah may conclude that Iran should hold onto its spent fuel to retain the option to reprocess. Since the Shah has acknowledged that Iran may build its own nuclear weapons, the country is not likely to turn all its spent fuel over to an international regime unless it can retain title and guarantee that the fuel will be returned on demand. In recent statements, the Shah has hinted that Iran may eventually be willing to establish a regional reprocessing plant on its own territory.

Israel

Although Israel has two research reactors, one supplied by France and the other by the U.S., little is known about Israel's nuclear power program. According to the Nuclear Proliferation Handbook, published in September 1977 by the U.S. House of Representatives' Committee on International Relations, Israel has no nuclear electric generating capacity. The same source claims, however, that by 1990 Israel will be producing 3.9 gigawatts and 5.0 gigawatts in conventional oil-fired plants. Given the widely held belief that it already possesses nuclear weapons, it seems extremely unlikely that the Israeli government would wish to contribute to an international spent fuel storage regime.

Pakistan

With virtually no fossil fuels and its hydroelectric potential fully tapped, Pakistan views nuclear energy favorably. It has a 125 MW natural uranium commercial reactor at Karachi, and a 600 MW plant under construction at Kundian that should be operative by 1982. A 1975 IAEA study of Pakistan's energy requirements recommended that two dozen 600 MW reactors be constructed by the year 2000. However, escalating capital costs will probably prevent Pakistan from meeting this target.

Meanwhile, Pakistan has concluded an agreement with France for the construction of a reprocessing plant at Chasma in north central Pakistan. Six 600 MW LWRs will be built at the same location. The U.S. has expressed concern about the proposed reprocessing facility, but construction will probably continue nonetheless. In September 1977, the French and Pakistani Foregin Ministers stated that France would honor the contract and supply the reprocessing plant to Pakistan. Pakistan's Foreign Minister dismissed the importance of U.S. pressure, and the suggestion that France was delaying delivery of blueprints and equipment, by saying that cooperation with France was proceeding according to the agreed timetable.

A Pakistani Embassy official in Washington has said that Pakistan will keep its own spent nuclear fuel if it proceeds with the reprocessing facility (although it could "keep" the fuel in a regional storage regime outside Pakistan by retaining title and the right to recall it on demand). The official also suggested that Pakistan might open its reprocessing facility to Iran and the Arab States of the Persian Gulf, if and when they decide to construct their own nuclear power reactors.

FUTURE POSSIBLE REGIONAL STORAGE SITES

None of the countries surveyed above may ever want to host a regional spent fuel storage facility. Nevertheless, it is possible to determine which nations make the most suitable candidates by using selection criteria developed by Dr. David Deese in Part I. The criteria are divided into national and international spheres. National criteria include a) the attitude towards nuclear proliferation, b) the internal political stability, c) the status of nuclear power programs and the degree of public opposition, d) the amount of aid the nuclear program and the economy would receive from the regional facility, and e) the geographic, geologic, hydrologic, climatic, and demographic suitability of the proposed spent fuel storage site.

Of the countries surveyed, Australia and Japan rank highest in their demonstrated encouragement of nonproliferation. More important, this encouragement is likely to continue. This is demonstrated by their diplomatic activities and their deliberate choice, despite having the necessary technology, not to develop nuclear weapons. In Australia's case, it is also demonstrated by the stringent safeguards that the government places on Australian uranium exports.

All the other states surveyed have formally committed themselves to nonproliferation. With the exception of the ASEAN countries, however, none can be counted upon not to develop nuclear weapons at some future time.

Security threats, real or imagined, abound. South Korea worries about North Korea; Taiwan, about the PRC; Pakistan, about India; Iraq, about Iran; Egypt, about Israel; and Iran, about the Soviet Union,

Afghanistan, and almost all the Arab states. All these countries may well want to build their own nuclear weapons in the near future. On a scale of one to ten, from unlikely to definite, the probability of these states going nuclear should be pegged between six and ten.

The potential nuclear energy-producing states within ASEAN – Indonesia and the Philippines – are not subject to the same strategic pressures. China is unlikely to threaten countries on its Southeast Asian rim, and, in any event, is too large for a nuclear deterrent to be effective. Vietnam could pose a military threat to ASEAN sometime in the future, but ASEAN sees the threat as neither immediate nor nuclear. The ASEAN countries will probably continue to hold the view that Vietnam has a long way to go to rehabilitate its economy; until the country has reconstructed itself, it can hardly pose a military threat to its noncommunist neighbors; and even if Vietnam did start to agress, it would do so through covert support for indigenous guerrilla movements in the ASEAN countries rather than through conventional invasion or nuclear attack. Given the absence of external military threats to ASEAN in the foreseeable future, these countries will have little or no reason to develop nuclear weapons. The likelihood of the Philippines or Indonesia desiring nuclear weapons, therefore, might rate two and four on the scale.

When internal political stability is considered, Australia and Japan again come out on top. Both are practicing democracies with predictable and stable electorates, a strong tradition of conservative civilian government, and little disruptive internal dissent or record of guerrilla activities. Both would rate one on the scale.

Countries ranking second under this criterion would include Taiwan, Indonesia, and Egypt. All have a relatively high degree of internal stability, combined with a low level of terrorism or organized internal antigovernment dissent. More important, the institutions of government in these countries would not necessarily dissolve into chaos or a succession of weak, irresolute military juntas if Premier Chiang Ching-Kuo, President Suharto, or President Sadat were to be suddently incapacitated. Many potential replacements with sufficient skill and confidence are ready to take over; and the governments, backed up by strong institutions, are secure enough to withstand decapitation without major dislocation.

The brittleness of one-man rule and shallowly-based government institutions in the remaining nations mean that unpredictable changes of leadership and policy could take place without much warning. In addition, several of these countries would have trouble guaranteeing the security of nuclear power plants and spent fuel facilities because of active antigovernment guerrilla organizations.

One of the most striking examples of this is the Philippines, with two excellent potential storage sites on the Bataan Peninsula and the island of Palawan. The Bataan site lacks security, geologic stability, and proximity to the megalopolis of Manila. However, the security is especially worrisome. Communist New Peoples Army (NPA) groups are active in the Peninsula and have posed a security headache for Philippine Constabulary units since the first Westinghouse PWR was sited in 1977. Palawan poses a different set of security problems. Although NPA and other guerrilla groups have never been active there, the Island, situated on the western side of the Visayas Region, headquarters the Western Command. A combined naval, air, and ground force established by President Marcos in 1975, the Western Command protects Philippine oil concessions in the South China Sea, and Philippine-occupied islands in the Paracels and Spratleys from possible encroachment by three other countries that claim them as their own – Vietnam, the People's Republic of China, and Taiwan. If hostilities should erupt over the possession of these islands, the contestants might regard Western Command headquarters at Puerto Princesa – located in the middle of the 80-mile-long Island – as a legitimate military target. A spent nuclear fuel facility situated nearby would be in obvious danger of suffering damage, with possibly disastrous ecological and environmental results. Even if the spent fuel facility were to be sited at either extremity of Palawan, the potential for military attack would still remain.

The countries with the least political stability are Iran, Pakistan, and the Republic of Korea: various leftist and rightist guerrilla groups have already sabotaged public works programs in Iran; tribal minority groups striving for autonomy have caused the Pakistani government recurrent security woes; and the Republic of Korea faces the possibility of further aggression from North Korea on the one hand, and the danger of internal subversion on the other.

A different hierarchy of advantages emerges under the third criterion, the status of nuclear power programs and environmental opposition. Public opinion in Australia would be highly-mobilized against hosting a nuclear spent fuel storage regime. Any Australian government would be committing political suicide to suggest that other countries' spent nuclear fuel could be stored in Australia. In Japan, the strength of the opposition is less clearly focused on the environmental issue, but it is virtually certain that no Japanese government would consider taking in other countries atomic wastes or spent fuel under any circumstances.

Several other countries surveyed maintain little or no vehicle for public debate. The leaders of Iran, Iraq, Libya, Pakistan, South Korea, and Taiwan face negligible environmental opposition. High levels of literacy, and the existence of small but sophisticated middle classes with environmental awareness would mandate that the governments of the ASEAN states, Egypt, and Israel be cautious before deciding to establish a regional spent fuel storage regime on their own territory.

One example is relevant here. Philippine Atomic Energy Commission head Librado Ibe has said that, despite the "constitutional authoritarianism" of martial law in the Philippines, a vociferous environmental group loosely-based at universities and other institutions in Manila could cause trouble if the government decided to construct a regional storage facility on Philippine territory. Environmentalists have already protested an iron ore sintering plant in Cagayan de Ore, Mindanao; and a copper smelter in San Juan, Batangas, was removed because of pressure from the Island of Negros. The authorities are not worried that nuclear fuel storage could stir up widespread opposition in itself, but that it could catalyze more general opposition to the Marcos regime. Similar constraints might affect the governments of Indonesia and Thailand. In fact, Thailand has already demonstrated that its antinuclear lobby was strong enough to persuade one government not to develop a nuclear power program.

The next national criterion is concerned with how much financial support a spent fuel storage facility could provide a country's nuclear power program and economy. Although the economics are far from definite, it is assumed that such a facility would earn considerable foreign exchange for the host state. All countries surveyed would naturally weigh this feature positively in deciding whether to construct such a facility. The countries to which the financial advantages would be most compelling, however, are those most worried about the large capital costs of installing their own nuclear reactors. The Philippines, Egypt, and Pakistan can least afford nuclear reactors because world prices for some of their traditional exports are uncharacteristically low. However, the monetary rewards of a regional spent fuel facility might tempt them. South Korea and Taiwan are better off financially, but they might also find a profitable regional facility attractive. Australia might conceivably desire the possible financial benefits of its own regional storage and, possibly, reprocessing regime, but domestic resistance would make the proposal politically unacceptable. Among the oil producing states surveyed, perhaps Indonesia and Iran would be interested in the financial rewards: Despite its oil revenues, Indonesia remains poor, with enormous development requirements; and Iran urgently needs to augment its oil income in order to carry out massive internal development programs.

The rankings change once again when geographic, geological, hydrologic, climatic, and demographic factors are considered. With a sparsely-settled, dry environment and with ancient granite structures relatively free from major tectonic disturbances, Australia must rank at the top of the list. Other surveyed countries with large sparcely-settled areas and dry conditions are Egypt, Libya, Iraq, Pakistan, and Iran, although the latter would not rank as high as the others because of its susceptibility to earthquakes. The countries of the Pacific Basin generally have dense, evenly-disbursed populations, very limited empty land areas, and – except for Indonesia – unstable, earthquake-prone geology. A suitable site might possibly be found somewhere on the relatively highly-populated Indonesian Islands of Sumatra, Borneo, or the Celebes.

International criteria suggested by Dr. Deese include a) other countries' perceptions of the host country's stability and acceptability, b) the host country's role in the international nuclear market, c) the host country's status regarding international nuclear organizations and nonproliferation treaties, and d) the geographic proximity and ease of access of contributing countries to the nuclear spent fuel storage site.

The first of the criteria assumes that similarity in social, cultural, religious, economic, and security outlooks would encourage states to cooperate in a spent fuel storage regime hosted by one of them. Thus,

Table 5.1. National Criteria Determining Suitability of Particular Countries As Hosts for International/Regional Facilities for Spent Factor Fuel Management

	Increasing Levels of Suitability		
	10 - 6	5 - 3	2 - 1
1. Host Country Attitude toward Nuclear Proliferation	South Korea, Taiwan, Pakistan, Iran, Iraq, Egypt	ASEAN States	Australia, Japan
2. Internal Political Stability	Philippines, Iran, Iraq, Pakistan, South Korea	Taiwan, Indonesia	Australia, Japan
3. Status of Nuclear Power Program and Opposition to it	Australia, Japan	ASEAN States	Iran, South Korea, Libya, Iraq
4. Economic Benefit of Regime to Host Country	Australia, Japan	South Korea, Taiwan, Indonesia, Iran, Iraq	Philippines, Egypt, Pakistan
5. Natural and Population Factors	Philippines, Indonesia, Japan, South Korea, Taiwan	Iran, Iraq	Australia, Egypt, Libya

there are three broad regional affinities among the countries surveyed – East Asia, the ASEAN countries and Australia, and the Middle East. However there are centrifugal forces that could militate against too-close association within these groups: Japan's great economic power and broad international interests might adversely affect its relationships with Taiwan and Korea; and the religious and historical differences between the Iranians and Arabs, and the ideological divisions between the conservative and radical Arab states could be unsettling.

Nevertheless, in terms of geographical proximity and the development of mutual economic interests, these alignments remain useful groupings around which to build the structures of international spent fuel.

The next criterion is concerned with the host country's role in international nuclear markets. Only Japan, Australia, Pakistan, and, possibly, Iran have present or planned roles in international nuclear markets: Japan is a possible enricher, fabricator, and reprocessor of nuclear fuel for other countries; Australia is a supplier of uranium oxide and might eventually supply enriched uranium; and Pakistan and Iran present possible reprocessing locations. If these roles are developed so that other countries become dependent on the services offered, the exporters' influence in determining the location and membership of spent fuel storage regimes might be increased.

It is probable that important new international nuclear agreements and nonproliferation mechanisms will develop before the start of any regional spent fuel regime. Thus, it is not appropriate to discuss the third criterion at this time. Present assessments might bear little relevance to future realities.

The final international criterion – ease of access to a spent fuel storage site – strongly suggests that at least two regional sites, one in the Pacific Basin area and the other in the Middle East, would be convenient. However, intraregional tensions, especially in the Middle East, might make it impossible to establish a spent fuel regime within any one country of a region that would be acceptable to all others in the region. One potential solution to this dilemma is the Indian Ocean spent fuel storage regime proposed by Onkar Marwah. Conceivably, this could become the outlet for the Mideast's spent fuel. An Indian Ocean site, though beset by its own problems, would provide a neutral venue where tensions between Iran and the Arab states, and between India and Pakistan could dissipate.

Similarly, a Pacific Ocean island, internationally-controlled and geographically-convenient, could serve as a spent nuclear fuel site for the present and future nuclear-power-generating countries of the Pacific Basin. The most intriguing possibility is a Micronesian island, possibly in the Solomons, where the U.S. has maintained missile and nuclear weapons testing facilities for a number of years. However, current independence negotiations between the U.S. and indigenous Micronesian leaders clouds the picture somewhat. Assessing the feasibility of Micronesia as a host for a spent fuel facility must await final resolution of Micronesian independence.

If independence is negotiated, Micronesia's new government or governments would decide whether to provide an island for regional fuel storage. Authorities in the Solomon group might oppose the idea, despite its financial attractions, because of nuclear contamination fears that grew out of American hydrogen bomb tests on the islands of Bikini and Eniwetok in the 1960s. The peoples of those islands were evacuated for many years, and returned home only recently. In fact, the inhabitants will soon be evacuated again because radiation has entered the food chain, a condition that will persist for many years to come, perhaps indefinitely.

Assuming that a suitable island can be found in the Pacific, either in Micronesia or elsewhere, would the littoral nuclear states contribute to a spent fuel regime established there? Available evidence suggests that Taiwan and the Republic of Korea would probably be willing to do so if they could withdraw their spent fuel for reprocessing upon request. Because it has indigenous reprocessing facilities and toll contracts with European reprocessors, Japan would probably be disinclined. Although they would prefer a regional regime within ASEAN, Indonesia and the

Philippines might agree to join if enough international pressure was brought to bear. However, the Philippines has intimated that it wants to establish a regional storage facility within its own territory. It remains to be seen whether they pursue this course.

6 Western Europe

Robert Gallucci

The transatlantic debate over plutonium fuels – when, if ever, they will be necessary and how they ought to be controlled if they are necessary – branches into virtually every segment of the nuclear fuel cycle. American efforts to promote the idea of an international spent fuel storage regime have grown out of the broader policy of postponing commercial reprocessing of spent fuel. Since 1975 and the first London Suppliers' Group meetings, U.S. representatives have argued that storing spent nuclear fuel is more rational than building a small, uneconomical reprocessing plant. Although the case was made to the suppliers, it was intended for the recipients, those emerging nuclear countries that had accumulated only insignificant amounts of spent fuel. The Americans point out that they had roughly 60 operating nuclear power plants, but not a single commercial reprocessing facility.The huge 1500 ton per year Barnwell reprocessing plant would make sense in the U.S. when it was finished, but very few other countries could support a facility that was economically viable. The argument was intended to discourage the export of small plants, such as those planned by the French for South Korea and Pakistan, and by the Germans for Brazil.

The American position on storing spent fuel began to shift at the end of the Ford Administration, and was significantly changed in the Carter Administration. As part of the new Administration's policy of a "pause" in the movement to plutonium fuels, the U.S. favored the suspension of all plans for reprocessing and the continued storage of spent fuel. Every other nation was asked to follow the same strategy, although the right to make independent judgments on the issue was acknowledged. About the same time that Washington began to advocate storage for all countries, it also began to stimulate discussion and study of an international spent-fuel-storage regime. No one specific proposal emerged, but the objectives and potential advantages of such a regime were fairly obvious. Assuming the regime would take physical form rather than merely providing for legal transfer of title, regional storage facilities could limit, if not eliminate, the stockpiling of nationally-controlled spent fuel. This would represent the first step in the extension of international authority over sensitive material in the nuclear fuel cycle – in this case, the plutonium in irradiated fuel. For some countries, an international or regional storage facility would be the answer to the technical and political problems of radioactive waste disposal. From the perspective of nonproliferation, a well-constructed regime would support a policy of either abandonment or deferral of reprocessing, while easing concern over the adequacy of safeguards and security. The spent-fuel-storage working group, one of eight in INFCE, is supposed to explore the concept in greater detail. Ultimately, its impact will depend a good deal on European reaction.

The response to an international spent fuel storage regime in Europe will vary among countries and will depend upon the form in which it is offered. There are many questions about spent fuel storage that are of concern to the Europeans. How would it impact upon the immediate problem of radioactive waste management and the licensing of nuclear power reactors? How could it affect plans for long-term waste disposal? Would it prejudice future reprocessing and recovery of plutonium; that is, would it guarantee reprocessing services or would it offer potential customers an alternative? Would storage and transportation costs be shared, offered on a fee basis, or structured to provide incentive for participation? Would siting authority over reprocessing, and disposition of the plutonium product give commercial advantage to one or more countries? Would participation be made a condition for nuclear cooperation between buyers and sellers of nuclear fuel and reactors?

The radioactive waste management issue is difficult to assess because it involves technical and political uncertainties. The short term question for Germany, Sweden, Japan, and Switzerland is whether participation in a storage regime would help them meet domestic demands to find an acceptable method for disposing of nuclear waste. Although the specific conditions differ in each case, in all four countries the future of the nuclear power program is being held hostage to a waste management solution. Three solutions are technically possible: Spent fuel could be immediately reprocessed and vitrified waste could be disposed of geologically; there could be interim storage, with eventual reprocessing and geological disposal of waste; or there could be interim storage with eventual geological disposal of whole fuel elements without reprocessing. Variations of those solutions are then possible by undertaking storage, or reprocessing, or geologic disposal inside or outside each state's borders.

Would shipping spent fuel to an international storage site meet the domestic legal and political requirements of a "solution" to the radioactive waste management problem? For Germany, which will account for one-quarter to one-third of the spent fuel generated in Europe by the end of the next decade, the answer is "maybe." The Germans currently plan to reprocess and store vitrified waste within their own borders. They also expect to use the recovered plutonium in a fast-breeder-reactor development program. If the spent fuel were to be shipped out of country and stored at an international site awaiting eventual disposal or if it were to be reprocessed and the vitrified waste disposed of by the reprocessor, the storage regime would presumably qualify as a technically-sound solution to the waste management problem. However, this regime might not provide a politically-sound solution. Domestic opinion, legal and local, might become convinced that spent fuel has to be reprocessed in order to manage radioactive waste. In that case, if the nuclear establishment wishes to maintain its plan to use plutonium fuels in the future, it would feel that it must do more than opt for the temporary solution of international storage; it would proceed with plans for a national reprocessing facility. The issue of whether participation in an international storage regime would legally prejudice future plans to develop plutonium breeder reactors would depend upon who had authority over the decision to reprocess spent fuel and release recovered plutonium; furthermore, there is no technical reason why interim storage for 20 years or more could not be followed by reprocessing. However, there is a question about whether waste management arguments offered in support of reprocessing in Germany in the past have created a political requirement that reprocessing proceed now in order to support a breeder option in the future.

The same situation does not exist in any other country. An international storage regime would relieve the immediate waste management problem in countries such as Sweden and Switzerland.

The second issue raised by an international spent fuel storage regime revolves around technical uncertainty over the most desirable form for final waste disposal – and the political consequences of the resolution of the technical questions. It is argued by some that ultimate disposal of high level waste from nuclear power reactors can be accomplished safely and economically only if spent fuel is reprocessed; high level waste can then be separated, reduced in volume, vitrified, and removed for final geologic disposal. The only other known alternative for the disposal of nuclear power wastes is to omit the reprocessing, reencapsulate the spent fuel elements directly, and remove them for the same geologic disposal as vitrified waste. Advocates of reprocessing assert that the vitrified waste from reprocessing is more likely to resist leaching from ground water and remain isolated from the biosphere than is the unreprocessed spent fuel, even though oxide fuel is itself in a ceramic form and encased in a zirconium tube.(2) The point is disputed and the technical argument has not been resolved.

The second part of the dilemma involves the more immediate problem of accumulations of spent fuel in nuclear-reactor storage ponds. It is not only that the capacity of existing storage ponds is nearing exhaustion, though that has become an issue for more than one of the world's utilities, but that the image of filled storage ponds is becoming a political liability. The public views them as evidence of the unresolved state of waste management. As yet, no country deposited high level wastes, in vitrified form or as spent fuel, in permanent geological storage. All high-level nuclear waste generated from nuclear weapons and nuclear energy programs in the Western countries remains contained in liquid form in metal holding tanks next to reprocessing plants, or in spent fuel elements under water in storage tanks next to reactors. No one would argue that either is acceptable as a permanent solution; the nuclear industry, the utilities, and governments must admit that, although the benefits of nuclear power have been enjoyed for over twenty years, a means for permanent, safe disposal of the highly toxic wastes generated throughout that period has yet to be demonstrated. The problem, of course, is that industry, utilities, and government must make that admission while arguing for licenses and public support for more nuclear power reactors. It is at this point that a commitment to build a reprocessing plant is politically useful, even if it may not be technically necessary. It permits the nuclear energy establishment to show movement toward a solution for waste disposal; and it maintains the option of a plutonium economy in the future. If spent fuel were to remain in pond storage, plutonium fuels would continue to be a future option, along with the alternative of direct disposal of fuel elements. More storage ponds rather than a reprocessing plant would then be built and without adversely affecting, and perhaps even simplifying, the waste management problem. Again, the benefits of a near term commitment to reprocessing are political.

Participation in an international spent fuel storage regime would probably relieve a country from at least the full responsibility of determining where and in what form its high level or spent fuel waste should be deposited for final storage. Therefore, the technical debate over vitrified waste versus direct spent fuel deposit is only relevant to a decision of whether to participate in a regime to the extent that participation would concede the potential viability of radioactive waste management without reprocessing. Thus far, initiatives for an international storage regime have left open the question of whether reprocessing would ultimately be undertaken before disposal. Advocates of reprocessing and plutonium fuels who endorsed the international storage regime could undercut the waste management rationale for reprocessing and give credibility to the option of interim storage followed by direct disposal. This is not likely to be decisive in decisions over participation; but the issue has already emerged in the INFCE discussions and it would be a consideration in some countries, especially Germany, France, and Britain.

A factor that is more likely to be decisive, at least for the French and the British, is whether the storage regime is perceived as guaranteeing future contracts for their national reprocessing plants, or whether it is offering potential customers an alternative to their reprocessing services, thus undercutting the services market. The underlying assumption here is that the international spent fuel storage site would be collocated with an existing reprocessing plant, or on a site where one is planned or could be located. This would have to be the case if eventual reprocessing was to remain an option, since the cost of shipping spent fuel twice would be prohibitive.(3)

Oxide reprocessing is currently underway at La Hague in France, and is planned, pending the outcome of a recently-completed inquiry, at Windscale in Great Britain. These are, therefore, two likely sites for storage. Mol, Belgium, the site of the old Eurochemic plant, is another. The yet-to-be-determined site of the planned German reprocessing plant could theoretically be a fourth were it not for the popular sentiment against accepting foreign waste in that country.

Because the French and British are in the business of reprocessing foreign fuel, they would almost certainly not support an international storage regime unless they were reasonably certain that reprocessing would eventually be undertaken either for the recovery of plutonium and uranium, or as a necessary step in waste disposal. The assurance of eventual reprocessing would have to be included in any agreement establishing a storage regime; and the U.S., Canada, and Australia would have to guarantee that they would not use their right to block the reprocessing of fuel originating in their countries. This would undoubtedly be a delicate issue in all three supplier countries; they would have to balance their desire to maintain control over the reprocessing of the fuel they supplied against their interest in seeing a storage regime established.

The answer to the question of who would pay for the transportation, construction, and operating costs of a facility has always been obvious to both the party proposing to store the fuel and the party providing the fuel to be stored: each argues it should be the other.(4) The U.S. Government may have set a precedent with its announcement in October 1977 that it was proposing to take title and physical possession of spent fuel from its domestic utilities for a one-time fee covering interim storage and permanent geological disposal. The issue of eventual reprocessing was neatly put aside with the following provision.

> If, at some time in the future, the U.S. should decide that commercial reprocessing or other energy recovery methods for spent fuel can be accomplished economically and without serious proliferation risks, the spent fuel could either be returned with an appropriate storage charge refund, or compensation could be provided for the net fuel value.(5)

The proposal may be precedent setting, but it should be expected that some will resist it. India is said to have argued that it should be compensated for the value of the plutonium and uranium in the spent fuel that the U.S. proposed to "buy back" because it would have recovered this material had the Americans permitted reprocessing in India. The idea of paying for the storage and disposal of the spent fuel apparently never occurred to New Delhi. Countries such as Belgium, Spain, and Italy, which may account for 15 to 20 percent of European spent fuel by the end of the next decade, could well take a position comparable to India's; all three countries have plans for a nuclear program that includes reprocessing. However, the most important countries in Europe, in terms of the volume of spent fuel generated, will probably find the American financing precedent acceptable in principle. This assumes that Britain and France, as nuclear-weapons states with commercial reprocessing ambitions, would provide storage sites and charge a fee rather than pay one. If Germany were able to reconcile participation with its past arguments in favor of reprocessing, the payment of a storage fee should not be an obstacle for its hard-pressed utilities. It is also possible that the U.S., together with other supplier countries, would subsidize the enterprise, either directly or through the IAEA, in order to provide an incentive for countries to participate.

The issue of commercial advantage is wrapped up in the prior questions of whether reprocessing would eventually be undertaken and at what cost to whom, and under what terms would the recovered plutonium be returned if it was reprocessed. In the first instance, European and non-European countries participating in an international spent fuel regime on a fee basis would have to be convinced that a supplier monopoly was not being created in the name of nonproliferation. This not only implies that the storage fees would be perceived as fair and reasonable – something the French reprocessing charges clearly are not – but also that access to recoverable plutonium in the event of eventual reprocessing be based on some principle of legitimate need rather than limited to the host country. Although the nonproliferation benefits of having recovered plutonium limited to one or two states are obvious, especially if they are the nuclear weapon states of France and Britain, it would hardly be acceptable to other countries if the French made use of the material in their own plutonium breeder programs. It would undoubtedly be difficult to reach an agreement on release criteria; and uranium credits would certainly be preferable to returning plutonium, no matter how it might be "denatured." Ultimately suspicions of commercial motivation for nonproliferation initiatives could run too high to expect the Spanish, Germans, Italians, Japanese, and others to accept permanent exclusion from plutonium fuels if the French and British proceeded to use them.

Finally, a concern of all potential participant countries would be whether the regime is structured to require participation rather than to invite it. Article XIIA of the IAEA Statute states the following:

> With respect to any Agency project, or other arrangement where the Agency is requested by the parties concerned to apply safeguards, the Agency shall have the following rights and responsibilities to the extent relevant to the project:...5...to require deposit with the Agency of any excess of any special fissionable material recovered or produced as a by-product over what is needed for (research or in reactors, existing or under construction) in order to prevent stockpiling of these materials, provided that thereafter at the request of the member or members concerned special fissionable materials so deposited with the Agency shall be returned promptly to the member or members concerned for use under the same provisions as stated above.

Under Article XIIA.5 of the Statue, it would be possible for suppliers of nuclear fuel to require a recipient country, as a condition of supply, to deposit "excess" spent fuel with the IAEA. If principal suppliers of nuclear fuel were to include such a requirement in their agreements for cooperation, comprehensive international spent fuel storage regime operated under IAEA auspices could be created fairly rapidly.

The amount of support that would be generated for required storage would very likely turn upon some of the factors mentioned above. In whom was the authority invested to determine the criteria for "excess"? Whether and when the fuel would be reprocessed, and what would become of the recovered plutonium would presumably be determined by the same authority. If a commission within the IAEA were the authority, its makeup would be critical. It is hard to imagine very many countries willingly accepting the principle of required storage unless the terms under which they could retain their own spent fuel, or have access to either it or the plutonium were non-discriminatory and consistent with plans for future nuclear energy options. The criteria would have to depend on the technical viability and level of development of the project for which the material was to be withheld or withdrawn from the regime. Establishing such criteria, and assessing individual cases according to them, might not present insurmountable problems if the question of the timing and circumstances of large scale reprocessing could be agreed upon.

Attempting to create a regime through regional storage would not be politically practical in Europe. User countries, such as Spain, Germany and Sweden, and host countries, Britain and France, would have to agree to the regulating criteria and character of the executive body. It is unimaginable that either of the latter states would support efforts to impose a regime, and without their support a regime would not stand. The assessment on this point, then, is that the virtues of a relatively-comprehensive, legally-binding regime with implicit sanctions attached are important enough to pursue the required storage concept; but realization of the regime would seem politically practical only if its advocates were willing to take a softer position on the circumstances under which reprocessing would be undertaken. In short, there is likely to be a trade-off between the inclusiveness of a storage regime and the height of the barriers to reprocessing and access to plutonium.

NOTES

(1) This chapter was researched and written by the author while on leave from the U.S. government. Although he is currently a member of the Policy Planning Staff of the Department of State, the views expressed in this chapter are his and do not necessarily reflect those of any agency of the U.S. government.

(2) Advocates of reprocessing also argue that it is possible to partition the waste so that highly-active fission product waste, which needs to be stored for hundreds of years, can be separated from the transuranic wastes, which require isolation for hundreds of thousands of years. Once separated, transuranic waste could be eliminated by the transmutation in power reactors. Critics note the difficulty experienced so far in efforts to partition and transmute waste; they argue that reprocessing ought not to permit this step until it is clear it can actually be taken on a commercial basis.

(3) The shipping cost within Europe is estimated to be roughly the same per ton of spent fuel as the cost of construction and maintenance of a storage facility. See the discussion of transportation and storage in the IAEA's Regional Nuclear Fuel Cycle Centers Summary Report, 1977.

(4) The uniformity and conviction with which these views were expressed in the course of interviews in Europe, the United States, and Asia were striking.

(5) "Department of Energy Announces New Spent Nuclear Fuel Policy," Organizing Conference of the International Nuclear Fuel Cycle Evaluation, Doc. 21, October 19, 1977. The announcement also said that the offer would be extended "to foreign users on a limited basis."

III

Feasibility of Managing Spent Fuel Internationally

7 Technical Considerations

Marvin Miller

INTRODUCTION

International management of spent nuclear reactor fuel has been proposed as a response to nonproliferation concerns.

Energy resource and waste management considerations have been advanced in supportof the argument for closing the fuel cycle via reprocessing of spent fuel. A spent fuel storage regime is seen as providing a viable alternative to immediate reprocessing, while retaining this option for a future time when the cost/benefit tradeoffs involved in insuring a long-term fission option via the fast breeder reactor becomes clearer.

Large amounts of spent fuel in national hands, especially fuel which has been out of core for a long period of time, is seen as an invitation to nuclear mischief via covert or overt seizure followed by reprocessing in a dedicated facility.

This report addresses some of the important technical issues involved in the implementation of a long-term, retrievable spent fuel storage regime. What is the state of the art as far as the different interim storage modes are concerned? What is the nature of the tension between retaining the option for eventual reprocessing and the option for permanent disposal? How long must spent fuel remain underwater before dry surface, near-surface, or geologic storage becomes feasible? What are the relevant accident, sabotage, and transportation consideration; and how do these differentiate between the storage modes? What impact would recent technical initiatives to upgrade IAEA safeguards on spent fuel stored in reactor water basins have on the nonproliferation rationale for international management? How high is the spent fuel radiation barrier to diversion and reprocessing as a function of time?

These questions are addressed in the following pages: various storage concepts are described technically; environmental impacts and transportation are discussed; tentative conclusions and recommendations are summarized; and the safeguards and radiation barrier issues are examined.

ALTERNATIVE SPENT FUEL STORAGE MODES

Figure 7.1 presents a generic view of the back end of the nuclear fuel cycle, illustrating the role of the various interim storage modes. As indicated, the basic uncertainty at this time is associated with the role of the geologic mode. At first glance, the idea of engineering this mode, which is normally associated with ultimate disposal of high level waste, in a manner compatible with long term retrievability seems very attractive. (Short term retrievability, for safety purposes, is always a design requirement.) However, there are important technical considerations which argue for reserving the geologic mode for disposal, not storage. Other uncertainties are indicated in Figure 7.1: How long must spent fuel remain under water before dry storage becomes feasible; and what are the packaging requirements associated with the different branches of the option space? Before discussing the different storage modes, it is appropriate to note that the new element in the concept of interim storage at the back end of the fuel cycle is the emphasis on spent fuel rather than the solidified, high-level radioactive waste. Most of the storage modes discussed here for the former application have already been considered for the latter,(1,2) and much of the analysis is applicable with some modification.

Storage Under Water

The handling and storage of radioactive materials under water is a standard method of operation in the nuclear industry. Spent fuel has been routinely stored in water basins for many years: the first reactor pool, associated with operations of the Manhattan District, was put into service in the U.S. in 1943; and the first commercial pools for storage of PWR, BWR, and HWR spent fuel were completed in 1957, 1960, and 1962, respectively. While the concept of extended water storage and delay in reprocessing of spent fuel is new, de facto, both spent fuel and experience in storing it have been accumulating over the years, both in the U.S. and abroad. In addition, extended water storage of solidified high level waste in canisters was considered by the National Academy of Sciences Panel on Engineered Storage and the AEC in 1974-75. However, bare spent fuel differs from canned high level waste with respect to potential corrosion mechanisms and criticality, and specific questions have been raised regarding the integrity of spent fuel in extended water storage. An assessment of these concerns has been undertaken by A.B. Johnson Jr. His reported results (3) and some related considerations are summarized below; but first, to supply some context, the current practice in spent fuel water storage should be reviewed briefly.

The typical spent fuel water basin is a rectangular sunken tub; its walls are composed of several feet of reinforced concrete lined with a water tight barrier, such as stainless steel or fiberglass. Spent LWR assemblies are housed upright in racks mounted on the bottom of the pool; appropriate rack spacing and construction provide assurance

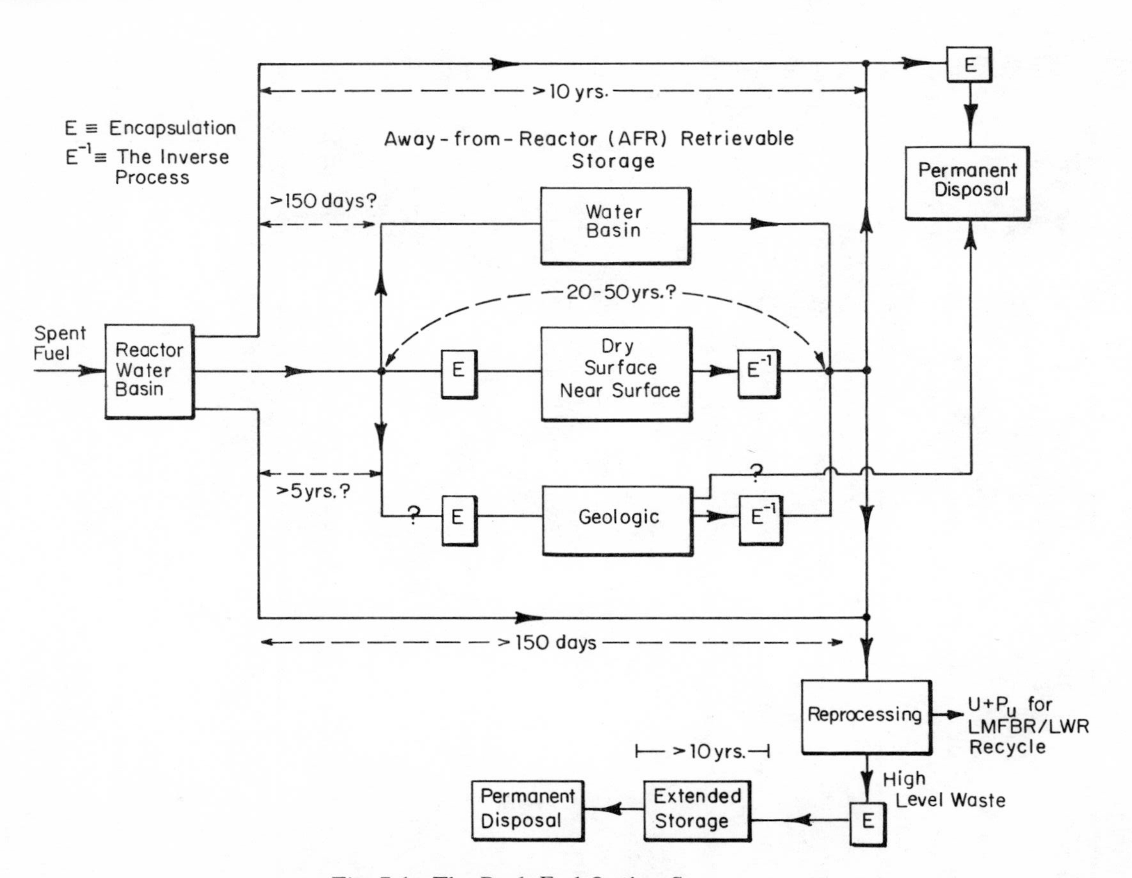

Fig. 7.1. The Back-End Option Space.

against accidental criticality. Criticality precautions are unnecessary for spent HWR fuel stored under ordinary – not heavy– water, and the short 0.5m fuel bundles are stacked in baskets; typically, there are 32 bundles per basket. The use of stainless steel, or boron-impregnated stainless steel instead, of aluminum racks permits roughly a factor of between 1.5 and 2 decrease in rack spacing, and hence a factor of between 2 and 4 increase in storage density. For example, the planned "reracking" of the U.S. Trojan PWR reactor pool, which had an original design capacity of 280 assemblies, will permit the storage of approximately 650 assemblies; this corresponds to an increase in capacity from 4/3 to 10/3 of a full core, with the normal annual discharge being 1/3 of a full core. The use of high density racks is an obvious, straightforward, but limited, solution to the growing shortage of space in reactor pools.

All spent fuel handling operations take place under water; these include the transfer from the reactor to the racks in the reactor pool, and any subsequent transfer to truck or rail casks for shipment to an Independent Spent Fuel Storage Installation (ISFSI). Cask loading in a reactor basin, and loading and unloading in a water ISFSI takes place in a separate pool, located next to the storage area and designed to withstand the accidental drop of a massive cask. Standard operating procedure also includes decontamination of incoming and outgoing casks before and after they are placed in the cask loading/unloading pool.

Adequate water pumping and heat exchanger capacity in a closed circuit system is provided to insure that the radioactive head generated by the fuel, which can be quite high for fuel recently out of core, does not cause the bulk water temperature to rise above $\sim 40^{o}C$. Radioactivity in storage pool waters is caused by the presence of fission products from the fuel, and by neutron activation products from crud deposited on the fuel during reactor operation. Filtration and ion exchange are the principal methods for controlling these species as well as other particulate and dissolved impurities, such as chloride ion. The water pool chemistry is usually sampled on a weekly basis, and monitors for airborne radioactivity operate continuously above the pools. The radiation dose rate at the water surface of current generation pools is less than 1 mrem/hr.

Regulatory guidance specifying requirements for design, site selection (including acceptable geology, meteorology, hydrology, and water supply), and physical protection of pool storage facilities is based, in the U.S., on USNRC Guides 3.24 and 1.13 for independent pools and pools at reactors, respectively.

The intuition that storage under water is a viable option for the long-term is usually inferred from the fact that the carefully-controlled pool environment is much more benign than that existing inside a reactor core where the fuel has been sitting for one to three years – the shorter/longer times are characteristic of HWR/LWR fuel. Hence fuel that survived reactor exposure without developing defects would be expected to age gracefully under water, while defective fuel which fails during storage would do so in a noncatastrophic manner and could be isolated from the pool water in closed canisters. Johnson's survey (3) of

U.S., Canadian, and European water storage experience and possible fuel and fuel cladding degradation mechanisms largely confirms this argument.

Both Zircalloy-clad and stainless steel clad uranium oxide fuel have been stored under water for long periods of time and subjected to visual inspections and radiation monitoring, with no evidence that the fuel bundle materials are degrading.(4) Observed fuel failure rates are low (~0.01 to 0.1 percent), and fuel assemblies with defective rods can usually be stored without special procedures. However, special equipment has been developed to handle failed fuel; "leakers" are contained in closed canisters.

Extrapolation of available experimental evidence suggests that known corrosion mechanisms, such as cladding oxidation, pose no threat to fuel integrity in water storage.

Oxidation of UO_2 to U_3O_8 at fuel defects occurs very slowly at pool temperatures. However, the reaction rate increases rapidly with temperature; and at temperatures of $\simeq 300^{\circ}C$ which may be attained in dry storage, substantial oxidation can occur. This assumes that an oxidant is available. The relevance of this to various dry storage options will be discussed.

To insure the credibility of water storage over the long-term, additional research is needed in several areas. It is important to determine the possible effects of pool temperature and water chemistry transients – such as might occur during loss-of-cooling capability accident – on the subsequent condition of stored fuel, the behavior of fuel defects as a function of defect type, cladding type and storage conditions, the effects of galvanic couples on the hydriding of zirconium alloys, and the definition of special effects such as crud layer environments and crevice corrosion. A low-level, selective spent fuel surveillance regime should be instituted as an integral part of this research program.

In conclusion, it appears that optimism regarding extended water storage is justified. The technology is well-established, and this mode is the logical reference case since water storage is the inevitable first step in any spent fuel regime. To be sure, no absolute assurance can be given that it is possible to store spent fuel retrievably, either under water or in a passive mode, for periods up to 100 years. Statements made at the recent Windscale inquiry to the effect that long term storage might lead to severe and costly deterioration problems cannot be dismissed out of hand, even though these statements were made in support of an application of British Nuclear Fuels, Ltd.(BNFL) to build an oxide reprocessing plant.

However, at least for the mid-term of 25 years, the weight of available scientific evidence and operational experience in water storage inspires confidence that immediate reprocessing is not a _technical_ requirement to insure ultimate retrievability of spent fuel.

Dry Storage

Dry storage of spent fuel becomes a viable option after it has cooled to the point where passive heat transfer from encapsulated fuel to its environment is efficient enough to insure that the fuel element temperatures are well below values which would lead to significant degradation over the long-term. Most of the known failure mechanisms are strongly temperature dependent. For example, as previously noted, recent experimental evidence (5) indicates that UO2 exposed at a cladding defect to temperatures $\simeq 300^{\circ}C$ will rapidly oxidize in air to U_3O_8, causing swelling and splitting of the cladding with exposure of more fuel, and ultimately release of finely powdered U_3O_8 and fission products. To avoid this situation, one can either restrict the out-of-core age of fuel/encapsulation method "space" to insure that fuel temperatures remain well below $300^{\circ}C$, or eliminate the risk of fuel oxidation by replacing the air atmosphere inside the canister with an inert environment. Current thinking, based on moving fuel from reactor basin to dry storage as soon as possible and simplifying the procedure for eventual reprocessing, favors the latter course with helium-filled canisters; helium also provides convenient leak detection. However, a view of dry storage as a de facto disposal mode would dictate different design requirements. This example illustrates the issues involved in engineering what is basically – as compared to water storage – a new, albeit straightforward, technology.

Three basic techniques for surface or near-surface storage are presently being considered in the U.S., Canada, and Western Europe: the dry well or caisson, the sealed cask, and the air-cooled vault.

The Dry Well or Caisson (Fig. 7.2)

In the dry well or caisson concept, canned fuel is stored below ground level in lined, vertical shafts which are sealed at the top for radiation shielding purposes. The rationale for this approach is that reliance on the soil for heat dispersion by conduction to the ground surface, for radiation attenuation, and for physical protection minimizes capital and operating costs; it provides a safe, economic alternative to long-term under water storage and the fuel is easily retrievable. On paper, this concept, developed over the past several years by the Atlantic Richfield Company and now being studied by the U.S. D.O.E. (6), looks quite attractive. The storage holes would be constructed as needed on a square grid; with a spacing of between 20 and 25 feet, storage of the spent fuel from 1 GWe-yr of LWR operation would require approximately one acre of land, assuming one PWR assembly per dry well. Heat transfer is by radiation and convection – primarily the former for young fuel – from the fuel to the can and from the can to the hole liner, and then by conduction through the soil to the atmosphere. The major design variables which determine the permissible decay heat load – and hence the age and amount of fuel which can be stored – are the soil conductivity, the hole spacing, and the hole

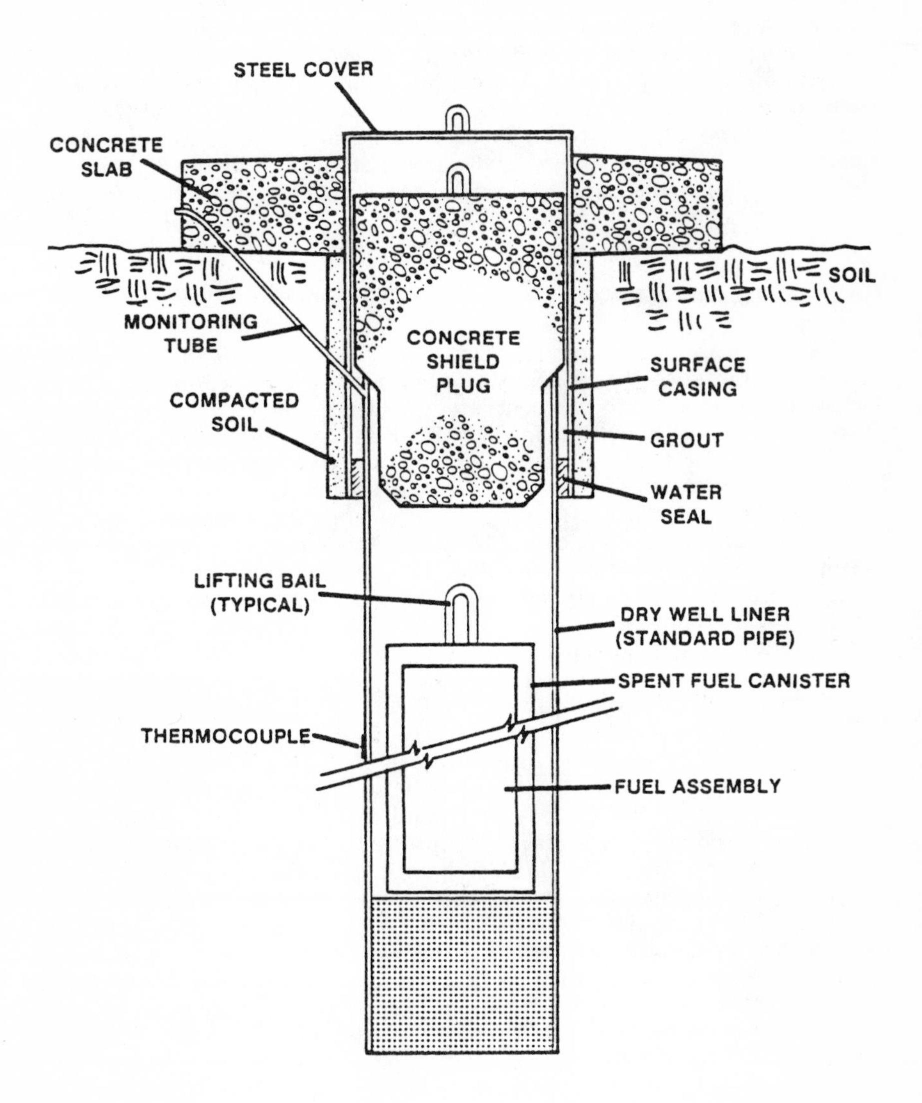

Fig. 7.2. Dry Well Concept.

diameter. Figure 7.3, taken from Reference 6, illustrates the strong dependence of the can temperature on the soil conductivity. It has been claimed (7) that spent fuel aged between two and three years under water can be stored in wells spaced 25 feet apart without exceeding a fuel cladding temperature of $\sim 380^{o}C \simeq 715^{o}F$, assuming a soil conductivity of 0.2 Btu/hr-ft-^{o}f. The indicated, allowable temperature is based on not exceeding two-thirds of the cladding rupture stress of PWR fuel rods.(6) However, this result is derived from computer heat transfer analysis and awaits experimental confirmation. The heavy reliance on the soil is both the basic attraction and a potential drawback of this storage concept.

The Sealed Cask (Fig. 7.4)

The storage cask method differs from the dry well in that fuel is stored in units which are self-shielding, cooling, and protecting. Each cask is a hollow reinforced concrete cylinder sitting on a concrete base. The spent fuel is contained in a steel can at the center of the cylinder, with possibly a thin lead layer between the fuel can and the concrete walls; lead is a much better shielding material than concrete, and its use reduces the thickness of concrete required. A similar design was recommended in the U.S. by the National Academy Panel on Engineered Storage as the optimum method for interim storage of high level radioactive waste; the dry well was not considered. In this application, natural draft air circulation through an annulus between the shield and the fuel can was necessary to remove the heat being generated at the design rate of 5 KW. However, because the heat rate per unit volume of aged spent fuel is lower than that of high level waste,(8) a sealed cask design seems feasible for the former. This route has been actively pursued at the Whiteshell Nuclear Research Establishment (WNRE) in Canada, where a development and demonstration program was started in 1974.(9) Tests at WNRE have verified that, at a design heat load of 2KW, corresponding to storage of 4.4MT of CANDU HWR fuel cooled for five years, the fuel cladding temperature is an acceptably low $200^{o}C$; similar heat load would be obtained from 1.6MT of five-year cooled LWR fuel. Besides fuel integrity, the other major potential materials problem associated with the cask concept is degradation of the concrete shield due to thermal stresses. At heat rates in the range 1.5 to 2KW, the high temperatures at the internal concrete wall result in tensile stresses on the outside surface which exceed the maximum tensile stress of the concrete. Under these conditions, surface cracking is predicted; however, it is expected that the reinforcing steel will prevent the cracks from growing to the point where the structural or shielding integrity of the concrete will be compromised. With the casks subjected to simulated freeze-thaw cycles, initial test have verified that surface cracking is minimal, even at higher-than-design heat loads of 5KW. The fact that heat transfer through the concrete was much better than predicted undoubtedly was a major factor in these results.

These results are encouraging, but more work is needed to insure the

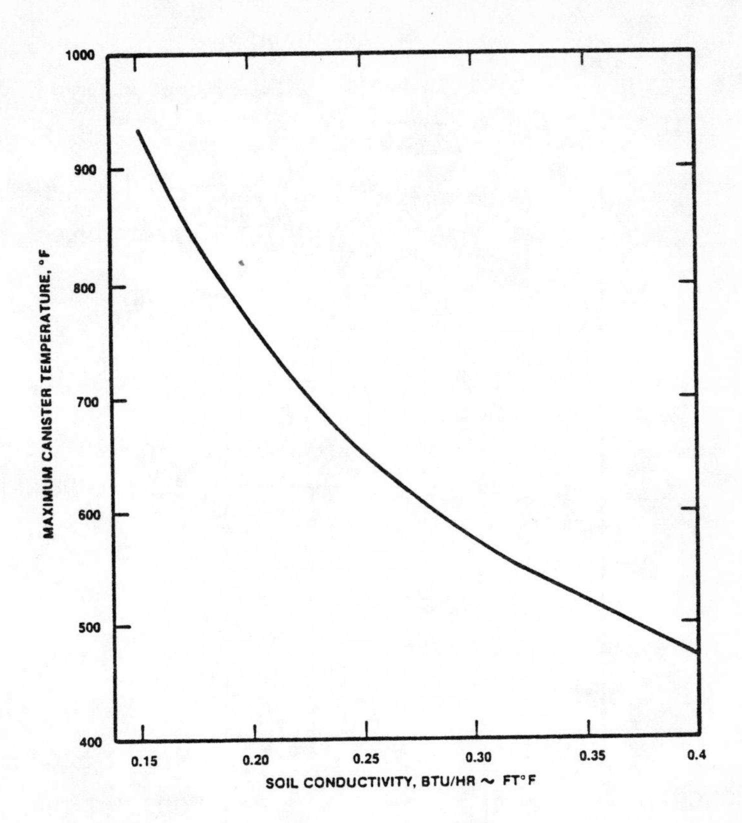

Assumptions:			
	Canister Thermal Power	=	0.964 Kw
	Canister Spacing	=	25 feet
	Canister Internal Diameter	=	13 inches
	Canister Atmosphere	=	Helium
	Canister Emissivity	=	0.3
	Fiel Rod Emissivity	=	0.4

Fig. 7.3. Maximum canister temperature versus soil conductivity for the dry well concept.

ability of the concrete to provide an effective shield for periods of up to 50 years. In this regard, treating the outer surface to increase its emissivity – thus lowering the heat load from the sun and increasing the permissible decay heat load – and to slow down weathering would be a major improvement.

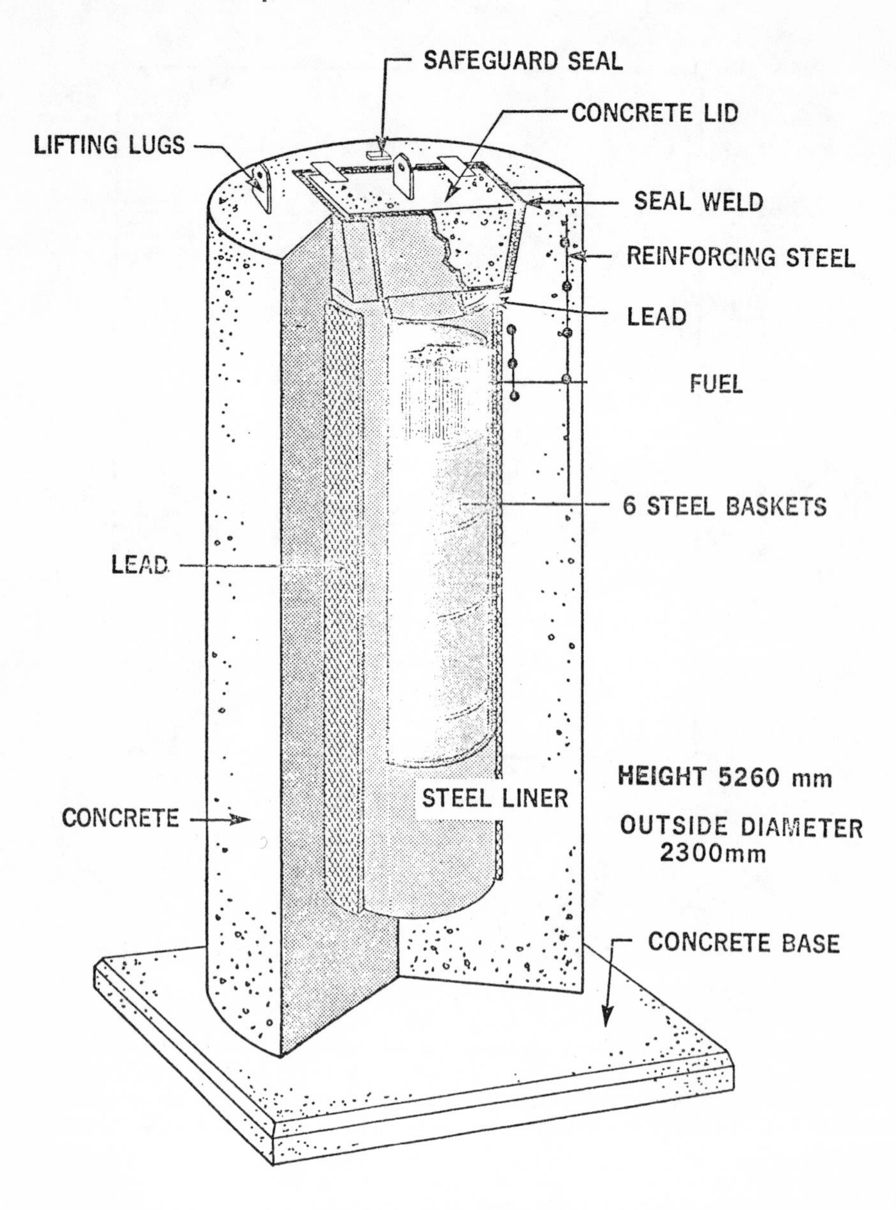

Fig. 7.4. Fuel Cannister.

The Air-Cooled Vault (Fig. 7.5)

In comparison to water cooling, both the dry well and the sealed cask are low density storage modes. The minimum spacing of sealed casks – on the order of 25 feet – is determined more by the need for access than heat transfer considerations. A dry storage concept with a packing density of the same order as water storage is the air-cooled vault. In this system, the fuel is contained within canisters and secondary canisters, or overpacks, and stored in closely-spaced vertical stacks in a large concrete bunker. This bunker is constructed partially below grade to reduce the radiation shielding and physical protection requirements. Heat removal occurs by the natural convection of air flowing directly through the vault, with the chimney effect of the hot air rising from the fuel providing the circulation. As with water storage and the concrete cask, the vault was previously considered in the U.S. for the storage of high level waste; it is now being reconsidered along with the dry well and concrete cask by the U.S. D.O.E. in their Spent Unprocessed Fuel (SURF) Facility Program. The convection valut concept has also been studied in Canada along with a variation, the conduction vault.(10) In both schemes, the CANDU fuel bundles are precast into zinc cylinders in an aluminum mold before canning to improve the containment and heat transfer. In the conduction vault variation, the fuel canisters are stacked tightly together and closed at the top by a finned aluminum shield plug. Heat flows up the canisters by conduction through the zinc castings, and is dissipated to the air by the shield plug fins. No cooling air enters the fuel area of the vault. This reduces the possibility that activity could become suspended in the exhaust air, at the "price" of less efficient cooling as compared with the convection vault. Important questions arise with regard to both schemes. Can the zinc be melted easily to allow recovery of the fuel? How much reduction in air flow can be tolerated? Is there any interaction between the zinc and zircaloy during casting or later in storage?

Geologic Retrievable Storage

Although the tension between reprocessing and ultimate disposal arises in all dry storage modes, it is felt most keenly in the case of geologic storage. On the one hand, it has long been appreciated that the plasticity and good thermal conductivity of salt make salt beds an attractive candidate medium for ultimate disposal of radioactive waste. On the other hand, the corrosive attack of brine tends to migrate up thermal gradients towards the heat source, and mechanical deformation of the salt at high temperatures demands a high degree of conservatism in packaging and thermal loading in order to keep the retrievability option open for an extended period of time. D.O.E. has recently estimated(11) that the thermal loading of a generic salt repository at a depth of 2000 feet must be restricted to 36KW/acre if 25-year retrievability is to be assured; keeping the retrievability option open

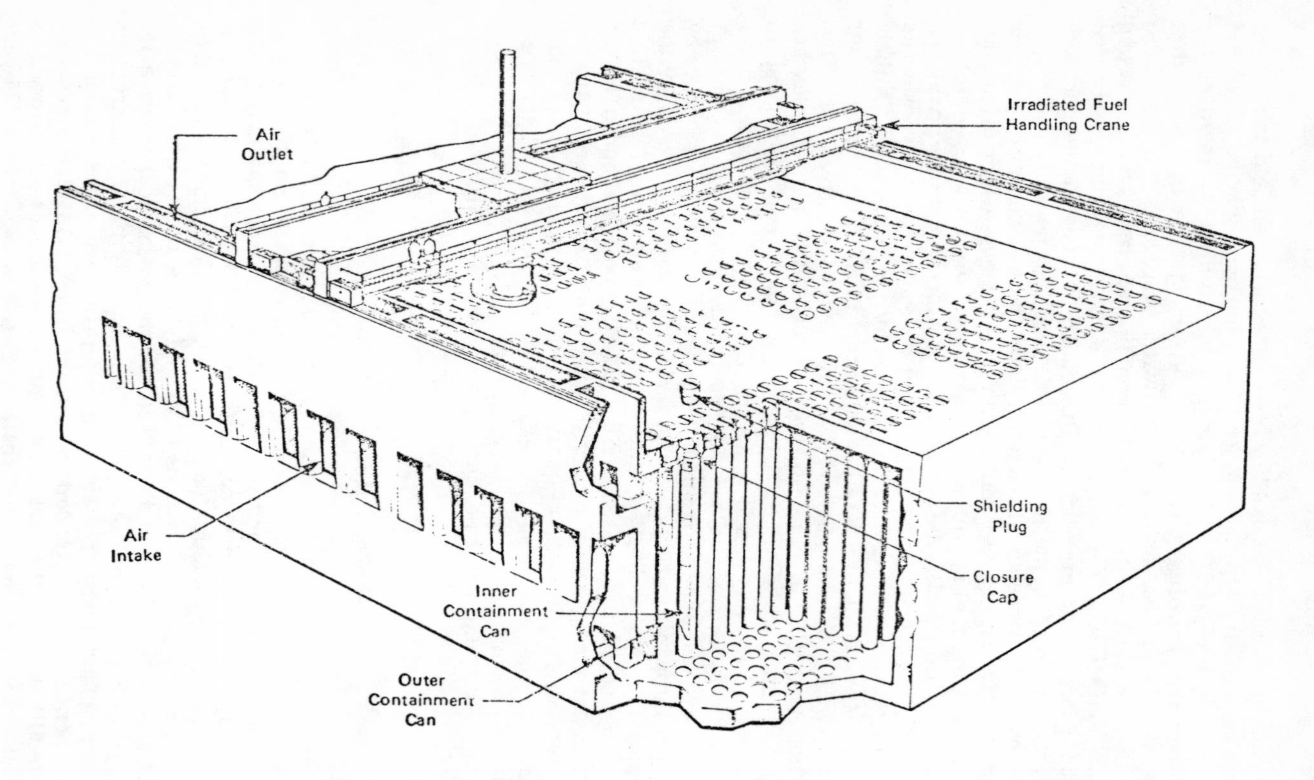

Fig. 7.5. Fuel Storage Convection Vault.

only for a five year initial "shakedown" period would allow densities of 150KW/acre. For non-salt formations, such as basalt, granite, and shale, corrosive attack is not known to be a problem but similar heat loading retrievability restrictions exist. However, current knowledge of canister-rock interactions is meager. Although it would be nice "to have your spent fuel and dispose of it too," the geologic mode is probably best reserved for ultimate disposal of either spent fuel or high level waste.

ENVIRONMENTAL IMPACT

The greatest potential hazard to the public from stored spent fuel is the release of radioactivity caused by missile impact or containment failure via loss of coolant. Even if the radioactive release in some accident, natural disaster, or sabotage attempt were limited, the retrievability option might still be significantly degraded and a large amount of fuel might have to be relocated in a short period of time. Given the relative state-of-the-art in water and dry storage, one would expect that statements relating to the environmental impact of the former would be much more definitive. This situation is reflected in the generic environmental statement recently issued by the US Nuclear Regulatory Commission.(12) Indeed, the discussion of normal and abnormal events and their consequences is entirely confined to water storage. Their basic conclusions can be summarized as follows:

1) Pool water and air quality can be easily monitored because there is essentially only one process stream, and the environmental impact of normal operation is nil. The volume of wastes associated with water cleanup is about $2m^3$/GWe-yr, with a maximum associated radioactivity of about 10 Ci/GWe-yr of beta-gamma activity.

2) The environmental impact of such events as fires, explosions, earthquakes, missile accidents and accidental criticality is small. This confidence is based primarily on the fact that the fuel sits below grade under a minimum of 12 feet of water, and is surrounded by walls of reinforced concrete which are typically six feet thick. Moreover, besides the basin itself, all important auxiliary equipment, such as fuel handling cranes and crane supports, is designed to prevent structures that could damage the fuel from collapsing into the pool.

3) A loss of cooling capability is potentially serious, but the large heat capacity of the pool water should provide adequate time for corrective action to be taken. This point can be illustrated with an idealized calometric calculation.

Assume that a 30 foot deep, 1,000 MT capacity storage pool contains 10^6 gallons of water and is loaded to capacity with fuel whose average heat rate is 10W/kg, corresponding to PWR fuel which has been out of core for one year. Then the total heat rate is

$$1000 \text{ MT} \times \frac{10\text{kw}}{\text{MT}} = 10^7 \frac{\text{joules}}{\text{sec}} \times \frac{3.6 \times 10^3 \text{ sec}}{\text{hr}}$$

$$= 3.6 \times 10^{10} \quad \frac{\text{joules}}{\text{hr.}} = 3.4 \times 10^7 \text{ Btu/hr,}$$

and the resultant temperature rise is

$$\frac{3.6 \times 10^{10} \frac{\text{joules}}{\text{hr}}}{\frac{4.18 \text{ joules}}{\text{gm}\ ^{o}\text{C}} \times \frac{3.77\times10^{3}\text{gm}}{\text{gallon}} \times 10^{6} \text{ gallons}} = 2.3^{o}\text{C/hr.}$$

Assuming the ambient water temperature was 40°C at the time cooling was lost, it would take roughly 60°C/2.3^{o}C/hr = 26 hours to reach boiling. To maintain the water level under boiling conditions would require that makeup water be supplied from an emergency source at a rate of

$$\frac{3.6\times10^{10} \frac{\text{joules}}{\text{hr}} \times \frac{1 \text{ hr}}{60 \text{ min}}}{\frac{540 \text{ calories}}{\text{gm}} \times \frac{4.19 \text{ joules}}{\text{calorie}} \times \frac{3.77 \times 10^{3} \text{ gm}}{\text{gallon}}} \cong \frac{70 \text{ gallons}}{\text{min}}$$

This is a modest requirement, and could be supplied from several sources, for example, the ultimate heat sink. If makeup was not supplied, the top of the fuel assemblies would begin to be exposed in about four days.

Environmental Summary

Almost by definition, a dispersed storage mode would be less affected by abnormal events of limited geographical scope. This would tend to favor the dry well and sealed cask as compared with pool storage and air-cooled vaults. Although low seismicity would be a site selection criterion for all modes, it would be of greater importance for the latter. The "other side of the dispersion coin" is that the task of surveillance and monitoring for evidence of containment deterioration and abnormal activity becomes more difficult.

As a corollary, canned spent fuel either in a hole in the ground, or surrounded by a thick concrete shield would be relatively immune to severe damage from tornadoes, earthquakes, airplane crashes, or saboteurs using conventional explosives. Even if the cask were to be toppled off its pedestal and cracked, the environmental impact would be small.

In addition to the fuel cladding, all dry storage modes rely on multiple barriers to contain possible radioactive releases. Nevertheless, more materials research and development is needed for all modes to assure the long-term reliability of these barriers; the possibility of accelerated rates of corrosion of metals or weathering of concrete under abnormal conditions must be taken into account.

Of the dry storage modes, loss of cooling capability, and criticality accidents would be more credible, and potentially more serious in the case of the convection vault.

TRANSPORTATION

An LWR shuts down annually to replace about one-fifth to one-third of its core. Storage of the discharged fuel at the reactor pool for a period of 150 days permits radiation levels and decay heat loads to diminish by approximately a factor of 100, as compared to levels immediately after discharge. This makes shipment feasible in massive casks – called "flasks" in England– that incorporate gamma and neutron shielding and provide cooling for the fuel. As with pool storage itself, shipping spent fuel in casks is not a new technology; present cask designs have evolved from experience gained since the mid-1940s in shipping fuel from commercial, military, and research reactors. There has been comparable experience in Europe, where about 590 MT of spent LWR fuel was shipped to reprocessing facilities between 1966-75.(13) Casks are usually classified according to the primary transport mode – truck or rail – and the nature of the coolant – water or air. Truck and rail casks now available or under construction weigh up to 35 MT/100 MT when fully loaded; This high fuel capacity substantially decreases the number of shipments and the loading and unloading required for a given amount of fuel. For example, the Nuclear Fuel Services NFS-4 truck cask weighs 22 MT, has a fuel capacity of 0.5 MT U, 1 PWR/2 BWR assemblies, and a two percent payload; the corresponding figures for the National Lead Industries NLI 10/24 rail cask are 97 MT, 4.7 MT U, 10 PWR/24 BWR assemblies, and five percent, respectively. Because only about one-half of the reactors in the U.S. have access to rail facilities, truck transport will continue to be an inevitable part of the spent fuel transportation picture. An aspect of the use of small capacity casks which is relevant to the rate of removal of spent fuel from basin storage is the typical cask turn-around time for shipping to an international storage facility. Assume that two NSF-4 type casks are available, and assume that shipment started 150 days after discharge: It would take approximately one day to load, decontaminate, seal, and check each cask before shipment from a PWR reactor pool; it would take about the same amount of time for similar operations at the other end; and about four days would be in transit. At this rate, it would take more than a year from the time of reactor discharge for 64 PWR spent fuel assemblies to have been shipped off the reactor site. The purpose of this example is to illustrate the potential magnitude of the "dead-time" transport problem. Not only does the fuel need someplace to go, but it takes some time to get there.

Related to the above is the question of cask availability. With the delay in reprocessing, the economic incentive for building casks has declined, and the number presently available or under construction in the U.S. and Western Europe is insufficient for shipment of large quantities of spent fuel. However, there is no reason to believe that –

given sufficient priority and economic incentive – fabrication of casks according to designs already licensed should be a major bottleneck. The NRC has estimated fabrication times of from 10 months to 3 years for a truck cask, and from 1.5 to 4 years for a rail cask. In order to receive a license to build a new cask, an applicant must demonstrate to the satisfaction of the Nuclear Regulatory Commission (NRC) and the Department of Transportation (DOT) that the cask provides the required containment, shielding, criticality control, and heat transfer under both normal and accident conditions. In particular, a detailed Safety Analysis Report (SAR) must be filed with the NRC to demonstrate compliance with the applicable code, 10 CFR Part 71; similar IAEA requirements are detailed in the "Regulations of Radioactive Materials, Safety Series No. 6." Because transportation accidents usually involve some combination of impact, puncture, fire, or submersion in water, the acceptance tests require evaluation of the cask and its contents for a 30-foot drop onto a completely unyielding surface, followed by a 40-inch drop onto a 6-inch diameter pin, followed by 30 minutes exposure to 1,475°F, followed by 24 hours of immersion in water. Although this is a formidable challenge, more extreme scenarios, some involving malevolent acts, can be imagined; and a breach of the cask containment with release of radioactivity near a highly-populated area could have serious consequences. Per vehicle mile estimates of the probability of accidents of varying severity and their consequences in terms of population radiation dose are derived in WASH-1238, "Environmental Survey of Transportation of Radioactive Materials To and From Nuclear Power Plants." As might be expected, the accident probability vs. consequence curves follow the pattern familiar from the Reactor Safety Study, WASH-1400; serious accidents occur via a series of improbable events, and hence have a very low overall probability. Although this is reassuring, it is important to emphasize that an international facility should be located in an area that could be reached with minimal shipping and that would avoid populated areas.

OBSERVATIONS

Extended spent fuel storage has become necessary because of the delay in reprocessing; and this delay has been sparked, especially in the U.S., by concern over weapons proliferation via the separated plutonium. Thus it seems fitting to conclude this brief overview with some remarks which focus on the nonproliferation implications of the technical aspects of spent fuel management.

1. Spent fuel which has been out of core for less than 100 years is still protected by a radiation barrier which necessitates remote handling, and hence is not as vulnerable as stockpiles of separated, decontaminated plutonium. The radiation barrier decreases sharply after this time due to the decay of Cs-137, the principal gamma emitter, which has a 30 half-life. However, the decrease in the level of the radiation barrier in time (Appendix

B, Table 2) might make diversion of spent fuel which is more than 5 years old somewhat easier.

2. Cooling spent fuel at the reactor basin for at least one year makes good sense in terms of utilization of at-reactor storage capacity, and in view of the problems associated with shipping intensely radioactive materials. Waiting for approximately five years would be even better. Smaller, less expensive casks could be used and there would be the option of moving directly from national wet storage to international/multinational passive dry storage, with a high degree of confidence in the long-term integrity of the stored fuel. Moving from wet to dry storage at an earlier time might be technically feasible, but it would require reoptimization to enhance heat transfer; this would increase the cost and might complicate retrievability if additional packaging is required.

3. The ability of at-reactor pool spent fuel surveillance to provide timely, unambiguous verification of attempted diversion should be significantly enhanced irrespective of the fate of initiatives for international storage. It is often argued that the political hurdles involved in instituting tougher safeguards are formidable; perhaps, but similar problems are involved in establishing an international/multinational spent fuel storage regime or bilateral spent fuel return arrangements. Some spent fuel will always be in national hands, and non-intrusive electronic surveillance techniques, while not foolproof, can make a major contribution to nonproliferation.

4. Despite a well-developed technology which permits high storage concentration with relative ease of access, water storage has the disadvantage of requiring active cooling and cleanup. The perpetual care of dry storage modes is less expensive, and they are less vulnerable in the event that supervision is lost for an extended period of time because of unstable political conditions or natural catastrophies. Hence, if water basins are chosen as the centralized storage mode, consideration should be given to locating them underground inorder to partially offset this liability. This should not involve great additional expense because they are currently built partially below ground.

5. Both the dry well and the concrete cask concepts are attractive as backups to water basins for interim storage of spent fuel or radioactive waste. Both require more development and testing.

6. Transportation of spent fuel should not be a major problem from the point of view of cask requirements, environmental impacts, or cost. However, multiple shipments with attendant rehandling increase all these factors; and public anxiety about the release of large amounts of activity via terrorist attack or accident makes

the optimization of interim storage logistics a priority item. The "obvious" solution is the colocation of interim storage with reprocessing facilities or with geologic formations suitable for disposal. The attractive feature of the latter is the potential for ready conversion to the former without additional transport, if a decision is made to close the fuel cycle in this manner. However, tying these concepts together would preclude early implementation of an extended storage regime because of the exacting technical site selection requirements for disposal. On the other hand, the viability of an extended storage regime would be compromised by locating it near an existing or planned reprocessing facility. These considerations, together with the political problem of finding suitable national sites for a multinational storage facility, have led to the suggestion(14) that a remote, sparsely-populated island would be a desirable site. It could serve initially for storage, and perhaps later for reprocessing and the production of methanol via fast reactors. The political and technical problems involved in finding appropriate locations have led some to characterize this concept as a "pie in the ocean." However, it should not be dismissed out of hand, if only because of the paucity of possible alternatives. O. Marwah presents an interesting discussion of the particular siting issues involved in the Indian Ocean area.(15)

APPENDIX A
SPENT FUEL SAFEGUARDS

The criteria for required levels of physical protection of nuclear materials under the IAEA regime is spelled out in INFCIRC/225, Rev. 1, June 1977. There are three categories – I, II, III – in order of decreasing stringency of safeguards. Both spent fuel and unirradiated natural or slightly enriched fresh fuel are in category III which sets conditions for use and storage within an area to which access is controlled. Special precautions for transportation include prior arrangements among sender, recipient, and carrier, and prior agreement between entities subject to the jurisdiction and regulation of supplier and recipient states, respectively. In the case of international transport, time, place, and procedures for transferring transport responsibility are to be specificed. Current surveillance requires that spent fuel pools be monitored by movie or video cameras every 15 to 30 minutes; it is presumed that fuel could not be covertly removed during this period. The video tapes are inspected every three months, and the movie film every six months. In addition, access to the pool area is monitored by gamma detectors. This is clearly not the ultimate in safeguards because it is possible to tamper with the transmission link between camera and tape recorder, and the inspection interval does not provide timely warning. Various technical initiatives are under development which, if implemented, would significantly increase the timely warning of attempts to divert fuel or tamper with the safeguard system. A prototype remote surveillance and interrogation system is now being tested. This system can be operated in various ways to monitor the status of a storage pool in almost "real-time."

For example, a TV camera takes a picture of the pool and stores it as a reference in a memory device. Every minute or so, the camera takes another picture and compares it with the one stored in memory. If something has changed, a status module on the camera transmits an alarm to an on-site multiplexer; from there it goes via telephone cable to a remote verification unit in, perhaps, Vienna. The primary technical problem is a high false alarm rate caused by the difficulty in discriminating between benign events – for example, changes in lighting over the pool – and alarm conditions. The same basic system could be used to send pictures of the pool in almost "real-time" by recording the video, digitizing it, compressing the bandwidth, and transmitting via telephone cable about a minute later. This technique is called "slow scan" video.

Using this same system, a fiber optic cable can be entwined through all the spent fuel assemblies and connected directly to a status module which senses continually in the cable, transmitting an alarm if there is a break in the on-site multiplexer.

The complete system, developed by Atlantic Research under contract to ACDA, consists of the monitoring units, on-site multiplexer, and remote verification unit; it can accept inputs from a variety of sensors. The motion detector is built by Fairchild Camera.

Special procedures have also been developed by Atomic Energy of Canada, Ltd. (AECL) and the IAEA for further protecting the on-power fueling feature of the safeguard system: fuel bundles are counted as they are being discharged from the reactor into the storage bay via the fuel transport system; cameras in the reactor building detect the removal of bundles from the reactor if it is done in any way other than via the fuel transport system; and a bundle radioactivity monitor verifies that the bundles in the storage bay have been irradiated and are not dummies.(2)

APPENDIX A NOTES

(1) For technical details, see draft report. "RECOVER – Remote Continual Verifications," Atlantic Research Corp., July 1977.

(2) A. Waligura et al., "Safeguarding On-Power Fueled Reactors – Instrumentation and Techniques, Paper IAEA – CN 36/185 presented at the IAEA International Conference on Nuclear Power and its Fuel Cycle, Salzburg, Austria, May 2-13, 1977.

APPENDIX B
THE SPENT FUEL RADIATION BARRIER

The attractive feature of spent fuel from the nonproliferation perspective is that access to the contained plutonium is inhibited by the intense gamma radiation field of the decaying fission products. How this barrier decays as a function of time after discharge can be illustrated by considering the properties of the spent fuel from a large 1150 MWe Westinghouse PWR of current design.(1)

1. Fuel Burnup: 33,000 MWD/MTU
2. Specific Power: 37.8 MW/MTU
3. Average Fresh Fuel Enrichment: 2.6% U-235
4. Square Fuel Assemblies
   a. Side Dimension: 21.4 cm
   b. Active Fuel Length: 366 cm
   c. Weight of Uranium: 520 kg.

Voluminous data on the nuclide concentrations and gamma decay energy from U-235 irradiated at a specified thermal neutron (2,200 m/sec) flux, ϕ, for a specified time, t, and then allowed to decay is available;(2) to extract the appropriate numbers, the given reactor chaacteristics are used to compute:

$$\frac{\text{U-235 atoms}}{\text{assembly}} = \frac{\frac{26\text{ kg}}{\text{MTU}} \times \frac{6 \times 10^{26}\text{ atoms}}{\text{kg mole}}}{235\text{ kg/kg mole}} \times \frac{0.52\text{ MTU}}{\text{assembly}}$$

$$= 3.46 \times 10^{25}$$

$$\phi = \frac{37.8\text{ MW/MTU} \times 3.2 \times 10^{16}\text{ fissions/sec/MW}}{6.66 \times 10^{25}\text{ atoms/MTU} \times 580 \times 10^{-24}\ \frac{\text{cm}^2}{\text{atom}}}$$

$$= 3.13 \times 10^{13}\text{ cm}^{-2}\text{ sec}^{-1}$$

$$t = \frac{33{,}000\text{ MWD/MTU}}{37.8\text{ MW/MTU}} = 873\text{ days} = 7.54 \times 10^{7}\text{ sec.}$$

From Fig. T-11, 11a of Reference 2, the gamma power in watts/assembly, S, can now be found for some representative times, t, after discharge:

Table 1

t	S
Discharge	15.2×10^4
1 month	17.3×10^3
150 days	3.5×10^3
1 year	12.1×10^2
5 years	208
10 years	138
30 years	121

To convert S into gamma flux, I, the assembly is modeled as a line source of length 366 cm. Then I in watts/cm2 one meter from the midplane of an assembly is given by

$$I = 2\int_0^{183\text{ cm}} \frac{S/366}{4\pi(x2 + 100^2)} dx = \frac{S \tan^{-1} 1.83}{2\pi \times 366 \times 100}$$

$$= 4.7 \times 10^{-6} \text{ S/cm}^2.$$

To take into account self-absorption by the 264 fuel pins in the assembly, the mass of the fuel is averaged over the entire assembly volume and the mass absorption coefficient characteristic of the 0.66 Mev fission product gammas from Cs-137 is used; this makes the principal contribution to the gamma activity after 150 days. This gives an average reduction in I of approximately a factor of 5. Thus:

$$I \rightarrow 0.94 \times 10^{-6} \text{ S/cm}^2.$$

Finally, to get the gamma dose corresponding to the foregoing gamma flux it is again assumed that all the fission product gammas have the effective energy of those from Cs-137, which have a mass absorption coefficient in water of 0.032 cm^2/gm. Since a dose of one rad represents absorption of 100 ergs/gm, the close rate D in rad/hr from a gamma flux I in watts/ cm^2 is

$$D(\text{rads/hr}) = \frac{I(\text{watts/cm}^2)\ 0.032(\text{cm}^2/\text{gm})\ 10^7(\frac{\text{erg}}{\text{watt/sec}}) \times 3.6 \times 10^3\ (\text{sec/hr})}{100\ (\text{ergs/gm rad})}$$

$$= 11.5 \times 10^6\ I\ (\text{watts/cm}^2)$$

$$= 10S\ (\text{watts/assembly}).$$

Hence, from Table 1, the gamma dose 1 meter from the midplane of the PWR assembly at time t after discharge is:

Table 2

t	D (rads/hr at 1 meter)
Discharge	1.5×10^6
1 month	1.7×10^5
150 days	3.5×10^4
1 year	1.2×10^4
5 years	2000
10 years	1400
30 years	1200

To get a feel for the barrier represented by these numbers, note that complete incapacitation begins at 10^4 - 2 x 10^4 rem – equivalent to rad for gamma rays at these energies. Exposure to about 500 rem will result in death for one-half of the individuals so exposed; this is the so-called LD 50 dose. Below about 200 rem, there are no discernible near-term effects. From the point of view of ease of commercial reprocessing via the Purex process, less shielding would be required for old fuel; there would be less of a problem with radiation degradation of the organic solvent tributyl phosphate; and it would be easier to commercial Pu decontamination levels because the hard-to-separate fission products zirconium and niobium would have decayed to insignificant levels. The relevance of these matters to reprocesisng in a dedicated facility is not clear.

APPENDIX B NOTES

(1) J.J. Duderstadt and L.J. Hamilton, Nuclear Reactor Analysis, John Wiley and Sons, 1976, p. 634.

(2) J.O. Blomeke and M.F. Todd, Uranium-235 Fission-Product Production as a Function of Thermal Neutron Flux, Irradiation Time, and Decay Time, Oak Ridge National Laboratory Report ORNL-2127, August 1957.

NOTES

(1) Management of Commercial High-Level and Transuranium-Contaminated Radioactive Waste, U.S. AEC environmental statement, WASH-1539, draft. Sept. 1974.

(2) Interim Storage of Solidified High-Level Radioactive Waste, Panel on Engineered Storage, National Academy of Sciences, 1975.

(3) A.B. Johnson Jr., Behavior of Spent Nuclear Fuel in Water Pool Storage, BNWL-2256, Battelle Pacific Northwest Laboratories, Richland, Wash., Sept. 1977.

(4) To be precise, some zircalloy-clad and stainless steel clad fuel has been stored satisfactorily for periods up to 18 years and 12 years, respectively; however, the bulk of the commercial high burnup LWR fuel and CANDU fuel currently in storage has been under water for five years or less.

A possible exception which bears further study is the evidence of intergranular corrosion of stainless steel-clad Advanced Gas Reactor (AGR) spent fuel which has been exposed in core to temperatures in the range of 450-600°C. The apparent cause is sensitization of the stainless steel at these high temperatures. No evidence of this phenomenon has been observed with stainless-clad LWR fuel which has been exposed in core to temperatures in the range of 280-340°C. (A steel in which chromium carbide has precipitated at the grain boundaries, thus decreasing the chromium content in these regions, is said to be "sensitized.")

(5) D.G. Boase and T.T. Vandergraab, "The Canadian Spent Fuel Storage Canister: Some Materials Aspects," Nucl. Tech., Vol. 32, pp. 60-71, Jan. 1977.

(6) K.H. Henry and D.A. Turner, Storage of Spent Unreprocessed Fuel (SURF), RHO-SA-40, Rockwell International, Richland, Wash., March 1978.

(7) L.M. Richards and M.J. Szulinski, "Subsurface Storage of Commercial Spent Fuel," American Nuclear Society Topical Meeting on the Back End of the LWR Fuel Cycle, Savannah, Ga., March 19-23, 1978, and M.J. Szulinski, private communication.

(8) Typically, high-level waste solidified in borosilicate glass with a fission product content of 20% would have a heat rate of 6W/kg at 10 years of age; the heat rate of spent LWR fuel of the same age is 1W/kg.

(9) M.M. Ohtu, The Concrete Cannister Program, AECL-5965, Feb. 1978.

(10) Report by the Committee Assessing Fuel Storage, Edited by W.W. Morgan, AECL 5959/1,2, Nov. 1977.

(11) Report of the Task Force for Review of Nuclear Waste Management, U.S. Department of Energy, DOE/ER-0004/D, February 1978, Appendix E, pp. 55-59.

(12) Handling and Storage of Spent Light Water Power Reactor Fuel, draft generic environmental impact statement, U.S. Nuclear Regulatory Commission, NUREG-0404, Vol. 1,2, March 1978.

(13) Regional Nuclear Fuel Cycle Centers, Vol. II, Basic Studies, 1977, Report of the IAEA Study Project, Section 14.

(14) The concept of an energy island incorporating spent fuel storage reprocessing, waste disposal, and fast breeder reactors has been extensively developed by W. Haefle and his associates. See, e.g., Nuclear Waste Storage and the Energy Island: A New Possibility for Action, International Institute for Applied Systems Analysis, Luxenburg, Austria, Oct. 1977.

(15) O. Marwah, Regional Storage of Spent Nuclear Fuel: The Indian Ocean Area, Program for Science and International Affairs, Harvard University, March 1978.

8 Economic Analysis

Boyce Greer
Mark Dalzell

INTRODUCTION

All countries that would consider entering into a multilateral arrangement for the storage of spent fuel would weigh the relative economic costs of various spent fuel storage programs. Therefore the economics of any arrangement should be acceptable to the countries that might participate. This chapter emphasizes the economic elements governing the acceptability of a spent fuel storage regime to the participating countries. As a first step, it is useful to determine if storage charges would compare favorably to national storage alternatives, including transportation costs to internationally-managed storage facilities. This is not a comprehensive comparison because the costs of international storage should also be compared to those of immediate or later reprocessing, which currently costs about $800/kg; but this comparison provides a necessary first step for decision-makers. The economic model of the spent fuel arrangement presented in this chapter shows that countries could be encouraged to form international regional arrangements for interim storage; these regional arrangements can be justified on economic grounds, even over great distances, and are likely to have broader political appeal than bilateral arrangements with nuclear supplier states.

ECONOMIC PROFILE OF THE BACK-END OF THE FUEL CYCLE

Before enumerating the various costs of pool storage, interim storage costs should be viewed in the perspective of total fuel cycle costs and the general costs of electricity production. Interim storage and transportation costs amount to only about three percent of fuel cycle costs, and the fuel cycle costs in turn constitute only 15 percent of the total generating costs. This means the interim storage and transportation costs are approximately 0.45 percent of total generating costs. Because the interim storage and transportation costs are so small when compared to the total generating costs of electricity, they will not affect any major energy policy decision to reprocess spent fuel, and consequent costs and benefits will not be sensitive to interim storage costs. This analysis does not consider interim storage costs with respect to these larger costs and benefits; only the relative costs of the various interim storage arrangements are considered.

Marvin Miller has described the various technologies available for the interim storage of spent fuel. These methods can be categorized into pool storage and dry storage. There is evidence that dry storage is considerably less expensive than pool storage – comparative costs per metric ton of spent fuel differ by a factor of 2 on the Hanson graph. Nevertheless, its total costs and benefits are less certain because the dry option has not been fully explored. The Canadians are an exception; they have performed extensive r&d on dry storage, and seem to prefer this alternative. However, because pool storage technology is the most developed to date and is currently in use in most countries with nuclear energy programs, this analysis will center on the costs of pool storage, both national and multinational.

The economic model which is developed in the next few pages divides the total costs of interim pool storage into capital, operating, and transportation costs. Capital costs depend primarily on pool storage capacity, and the specific design features of both the pool and auxiliary service facilities. Operating costs depend on staff requirements and maintenance costs. Capital and operating costs also depend on other business parameters, such as the form of financing, pool utilization, duration of storage, amortization period, and profit goals.(1) This analysis combines capital costs, operating costs, and the business parameters in order to simplify both the comparison of alternative storage programs, and the integration of storage and operating costs with transportation costs.

Transportation costs depend on the mode of transport and the distance traveled by the spent fuel; they can be expected to play an important role in a country's deciding whether to participate in a multilateral arrangement for interim storage of spent nuclear fuel. The largest portion of the transportation costs goes toward flask rental and shipping. A standardized heavy flask for rail barge shipment with a capacity of 5 metric tons of uranium (5-MT U) costs approximately $\$2x10^6$ which corresponds to an average daily flask-rental rate of about \$3000 per day.(2) This daily rental rate is based on an average flask utilization factor of about 80 percent. Obviously, there will be a large difference between the transportation costs of a national interim storage program and those of an international storage program that includes long transport routes. A multilateral program might include overseas shipment which would involve not only longer time periods, but also more complicated handling. Typically, intrastate transport costs are about \$30/kg-U, and the range of international transport costs is estimated at \$80-120/kg-U. Shipping costs predominate in intrastate transport, while flask rentals are the dominant costs in international transport.(3)

Because these cost components carry different weights in the two types of programs, the international transport costs will vary greatly depending on which component is optimized. Asit turns out, transport costs are about twice as high if shipping costs are optimized than they would be if flask rental costs were optimized.(4) Therefore, efficiency of flask use is an important consideration in all spent fuel transport. A flask utilization factor of 80 percent is a moderate estimate of flask-use optimization for calculating international transport costs. The following analysis also incorporates the additional transport cost components of insurance, permits, handling, managerial services, and port fees.(5)

THE ECONOMIC MODEL

The economic model of interim spent fuel storage proposed herein is based on several independent parameters. When computing the costs of the different storage programs, these parameters appear as independent variables in the cost equations. Before the mathematical relationships of the parameters are discussed, the parameters are defined and their places in the model are described. Six variables are considered: the size of the storage facility, the receipt rate of spent fuel at a storage facility, the annual production of spent fuel by a country of interest, the length of time that fuel is stored, the discount rate, and the unilateral transport cost.

Size of Storage Facility

An interim spent fuel storage facility consists of a large storage pool and the accompanying service equipment. The pool is lined with stainless steel and rests on a single slab of concrete, usually below surface level. The size of a storage facility refers to the pool's metric ton storage capacity of spent fuel. The service equipment includes a water cooling system to keep temperatures below 50°C, a pool purification system which maintains the purity and clarity of the water, a heating system, a ventilation and air-conditioning system, and a loading area for the transferral of spent fuel from transport flasks. The IAEA maintains that the size of an interim spent fuel storage facility is limited to 5000MT by the seismic integrity of the supporting concrete slab. One facility may consist of several cooling pools on the same concrete base. The size of a storage facility probably ranges between 300MT and 5000MT.

The IAEA lists the investment costs of a spent fuel storage facility as a function of the size of the facility. Thus, when the total costs of a storage facility are computed, the investment costs vary as a function of the size of the facility. The IAEA lists a range of investment costs for each size of storage facility. The middle value in each cost range has been chosen to give the following estimates for investment costs:

Storage Facility Investment Cost

Size of facility (MT)	Investment Cost ($\$x10^6$)
350	30
750	45
1000	60
2000	105
3000	150
4000	180
5000	210

The cost estimates for a particular size storage facility apply whether the facility is part of a unilateral or multilateral program. The IAEA cost estimates seem to be reasonable figures; the U.S. Nuclear Regulatory Commission has recently published similar estimates.

Finally, the size of the storage facility has a particular effect on the total cost of the facility. Because a facility must be constructed in segments, there is a non-utilization factor that increases the yearly cost to store a metric ton of spent fuel. Although the entire facility must be built before the program can begin, it will not be fully utilized until later. Therefore, a portion of the pool lies unused for a time, and this raises the unit cost of storing the spent fuel.

Receipt Rate of Spent Fuel

The term "receipt rate" refers to the number of metric tons that an interim spent fuel storage facility processes yearly. Although there may be, in fact, a correlation between the size of the facility and the amount of the fuel that it receives, this functional dependence is not a necessary condition. Thus, the model assumes that the receipt rate is not dependent on the size of the facility, and is therefore, an independent variable. Furthermore, the IAEA states that operating costs are more dependent on the fuel receipt rate than on the facility size; this relationship between operating costs and fuel receipt rate is incorporated into the model. It is also assumed that there is a uniform flow of spent fuel into the storage facility.

Annual Output of Spent Fuel

The annual output of spent fuel is the total reactor discharge of a single country. The size of most light water reactors, especially those under construction or on order, is about 1000MWe; this size reactor discharges approximately 30MT of spent fuel yearly. Thus, countries with small nuclear programs might have a spent fuel output of 50 MT/yr, while the reactors of countries with large nuclear programs, such as the United States, might discharge 1500MT/yr. Although the

IAEA calculates operating costs for a storage facility as a function of the receipt rate, this model alters that relationship somewhat. If a country puts all its spent fuel into interim storage, the annual output of spent fuel equals the receipt rate of the storage program. Thus, the model assumes that a country's total output of spent fuel goes into interim storage. This assumption allows the operating costs of spent fuel storage to vary as a function of the annual output of spent fuel. The following table which shows operating costs for a storage facility is extrapolated from the IAEA operating cost estimates:

Annual output s.f. (MY/YR)	Operating Costs ($/YRx$10^6$)
50	3.5
100	3.6
150	3.7
200	3.8
250	3.9
300	4.0
500	4.4
1000	5.2
1500	6.0

Length of Storage Program

The length of the spent fuel storage program is the number of years which the spent fuel is to be stored. Various economic and political factors set the likely length of a storage program to be between 10 and 20 years. International negotiations and technological progress indicate that the decision to either dispose or reprocess the spent fuel can be expected within the next decade. Thus, a 10 year program and a 20 year program are calculated in the model, but numbers can be generated for any length of time. The length of the storage program is an important element because it not only affects the total operating costs, but also the calculation of discounted present value of total program costs.

Discount Rate

The discount rate is necessary to compute the present value of future costs and investments. Given the discount rate, future investment and operating costs can be calculated back to their present value to get the total discounted cost of spent fuel programs. There are several conditions on which these types of calculations are based in the model: First, the model uses costs figured in 1976 dollars; second, a discount rate of 10 percent applied to all countries, which seems reasonable if we assume 60 percent debt and 40 percent equity financing of the projects, proportions typical of U.S. utility financing; finally, the model operates on the assumption that investment costs are lump sum cost flows and operating costs are uniform cost flows.

Unilateral Transportation Cost

In a national storage program, unilateral transportation cost is defined as the cost of transporting spent fuel from the reactor to the storage facility. The IAEA states that these costs range between $25 and $35 per kilogram of uranium. This model, however, takes a $10 to $40 range of internal transportation costs. Usually, these costs are generated by road and rail shipments, and remain small as a result of easy handling and the short distances travelled.

Unilateral transportation costs must be included in the cost comparison between national and multilateral storage programs. From this comparison, a breakeven cost is calculated for multilateral transportation; this determines whether a country would benefit financially from participation in a multilateral storage program, or whether it should build its own national program. If the multilateral transportation cost is greater than the breakeven cost, then, of course, it is less expensive for the country to conduct its own national program.

From these variables, storage costs can be computed (Tables 8.1-8.3). However, several other factors are calculated in the final cost estimates. For instance, given the length of a program, the annual output of spent fuel, and the facility size, the number of storage facilities needed in the program, the time that it takes to load the facility to capacity, and the average facility utilization factor can be determined.

A spent fuel program might consist of one or several storage facilities. If the length of the spent fuel storage program (Y) is assumed to be 10 years, then the number of storage facilities (n) depends on the annual output of spent fuel (A), and the size of storage facility (S). The algebraic relation of these four variables is:

$$n = \frac{A x Y}{S} .$$

When n is not an integer, the smallest integer greater than n is used in the calculations. In other words, when a country chooses the size of a storage facility to use in its storage program, it is assumed that it will build enough complete facilities of that size to accommodate the spent fuel accumulated over 10 years. It is necessary to know the number of facilities in order to compute the unilateral costs of a spent fuel program.

Another quantity which is necessary for the analysis is the length of time required to load an interim storage facility. This time period depends on the size of the storage facility and the annual output of spent fuel, and can be expressed in the following equation:

$$\text{loading time (YR)} = \frac{S\ (MT)}{A\ (MT/YR)} .$$

This quantity is necessary to compute the total operating costs of a spent fuel storage program.

Furthermore, assuming the annual output of spent fuel is constant, the average utilization factor of a facility can be calculated. While the facility is loading, there is unused space until the facility is fully loaded. This non-utilized space raises the average cost of storing a metric ton of spent fuel for one year. If only the utilized space is considered, the yearly storage cost can be ascertained. In this model, it is assumed that the facilities are loaded to capacity, although regulations might require emergency storage space in a spent fuel storage facility, and that would increase yearly storage costs per metric ton. Although the calculation of the utilization factor is more complicated, this factor also is a function of the annual output of spent fuel (A), and the length of the storage program (Y), with dimensions MT-YR:

$$U = \sum_{K=0}^{Y-1} A(Y-k) .$$

In Table 8.1, this utilization factor is used along with the other variables to calculate the total investment and operating costs of a particular unilateral storage program. The investment and operating cost estimates used were given by the International Atomic Energy Agency, and the stream of costs over the length of a program is discounted to get both the total present value investment costs (C_i) and operating costs (C_o). These investment and operating costs are dollar figures which, when added together, represent the present value of the total cost of a spent fuel program. This number, however, is less meaningful than a present value cost rate for storing a metric ton of spent fuel for one year; the present value cost rate reveals the amount of fuel storage received per dollar. Thus, total cost is divided by the utilization factor (U) to get the cost rate, whose dimensions are \$/MT-yr. (Storage cost rate= $\frac{C_i + C_o \ (\$)}{U \ (MT - YR)}$

This yearly cost rate per metric ton of spent fuel appears in Table 1 as a function of the annual output of spent fuel and the size of a storage facility. The cost rate estimates appear for both 10-year and 20-year storage programs. These cost estimates are in turn used to calculate the values in Table 8.2.

The cost values on Table 8.2 represent the least cost rate to any country participating in a multilateral storage program with a certain receipt rate. These cost rate estimates are derived from the least cost figures for a specified annual output in Table 8.1, and are also in the same dimensions (\$/MT-YR). Table 8.2 assumes that a multilateral storage facility with a specified receipt rate has the same operating and investment costs as a unilateral program of the same size and receipt rate.

The cost estimates on Table 8.3 represent the breakeven cost for transportation in an international spent fuel storage regime. Because the cost has been computed for both unilateral and multilateral arrangements, this final Table shows the point at which international transportation costs become so expensive that a multilateral arrangement is no longer desirable. Table 8.3 combines results from Tables 8.1 and 8.2 to arrive at breakeven transportation costs as a function of a country's annual output of spent fuel and the internal transportation costs for a unilateral storage program. The breakeven transportation cost estimates (T_{be}) are calculated from the least expensive unilateral program (C_u), the least expensive multilateral program (C_m), the annual output of spent fuel(A), and the unilateral transport cost (T_u).

$$T_{be} = \frac{Cu - Cm}{AxDy} + T_u$$ with Dy= discount factor for a uniform cost flow over y years.

The first trend to be noted in Table 8.1 is that, as the annual output of spent fuel becomes larger, the cost to store spent fuel becomes smaller for any given facility size. This is a general economy of scale with respect to the annual output of spent fuel. However, the most important trend to be seen in Table 8.1 is that countries with small nuclear programs, hence with small annual outputs of spent fuel, experience an initial decrease in cost, followed by a steady increase as the size of facility gets larger. For instance, in the case of a country with a 200MT output, the last expensive facility size is a 1000 MT, the unilateral storage cost for larger facilities gets increasingly more expensive.

This decreasing and increasing pattern in the colums of Table 8.1 is caused by two cost forces. Initially, the force which predominates and is responsible for the decreasing trend is an economy of scale; as the facility gets larger, the storage space for spent fuel becomes less expensive. After a certain point, however, another cost force predominates and the costs begin to rise. This second cost force is the non-utilization cost, which outweighs the economies of scale for large facilities (Fig. 8.1).

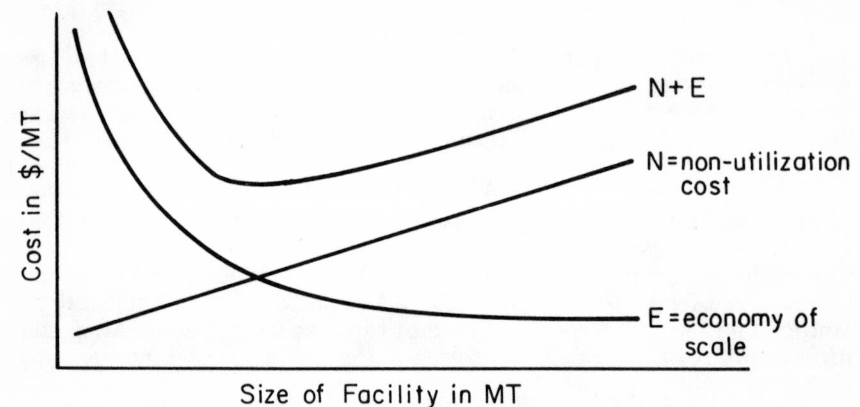

Fig. 8.1.

Table 8.1: Unilateral Storage Cost in \$/MT – YR

Case 1: 10 yr. program

Size of Storage Facility in MT	Annual Output of Spent Fuel in MT/YR								
	50	100	150	200	250	300	500	1000	1500
350	25109	16585	15269	13561	13377	12564	11770	10945	10671
750	24966	16609	11872	11288	10884	9542	8849	8139	7693
1000	30420	15333	14156	11176	11296	9847	8806	8012	7750
2000	46784	23515	15758	11880	13115	11373	9288	7398	7297
3000	63147	31696	21213	15971	12826	10729	9614	8111	6928
4000	74056	37151	24849	18698	15008	12547	10680	7673	6695
5000	84966	42605	28485	21425	17189	14365	8717	6828	6238

Case 2: 20 yr. program

	50	100	150	200	250	300	500	1000	1500
350	8197	5961	5230	4868	4652	4508	4179	3957	3869
750	8433	5309	4308	3994	3676	3462	3174	2887	2791
1000	8835	5564	4548	4055	3765	3573	3195	2907	2812
2000	13121	6605	5368	4311	4063	3611	3078	2684	2556
3000	17407	8748	5861	5273	4463	3893	3339	2716	2514
4000	20264	10176	6814	5133	4870	4253	3279	2533	2391
5000	23121	11605	7766	5847	4695	4608	3163	2477	2263

When the cost functions N and E are added together, the result is a total cost function with a minimum point. This is the phenomenon that is present in most of the columns of Table 8.1.(6)

If the unilateral costs in Table 8.1 are compared to the multilateral costs in Table 8.2, it can be seen that in both cases the unilateral costs are much higher than the multilateral costs. Table 8.2 shows the storage cost rate in $/MT as a function of the receipt rate of spent fuel for a multilateral facility. These cost rates apply to any country participating in a multilateral program, regardless of the size of its annual spent fuel output. For instance, if country X and country Y, with 50 MT and 250 MT annual spent fuel output, respectively, participate in an international storage program with a total receipt rate of 500 MT, both will pay $8,717 to store a metric ton of spent fuel for one year. The total cost for X and Y to store their output for one year is found by multiplying their annual output of spent fuel by the cost rate.

Table 8.2. Multilateral Storage Cost in $/MT-YR

	10 yr. program	20 yr. program
100	15333	5309
150	11872	4308
200	11176	3994
250	10884	3676
300	9542	3462
500	8717	3078
1000	6828	2477
1500	6238	2263
2000	5913	2146
3000	5586	2027
4000	5428	1969
5000	5333	1935

The multilateral situation is most easily observed in Fig. 8.2, where the cost rates for both cases are plotted as a function of receipt rate. The cost rate becomes relatively stable for a 10-year storage program at a receipt rate of about 1000MT/yr, while the cost rate levels out much sooner for a 20-year program, about 500MT/yr. The cost function for a multilateral program of an intermediate length lies between these two cost functions.

Table 8.3 lists the breakeven transportation costs for both cases. The most important trend in this Table is the gradual decline in breakeven transportation cost as the annual output of spent fuel gets larger. Because of this, countries with large nuclear energy programs will find it hard to justify participating in a multilateral program that involves complicated and expensive shipping arrangements. For instance, a country that produces 1000MT of spent fuel and faces a $20/kg shipping cost, must participate in a multilateral program which involves transportation costs of less than $32/kg in order to gain economically. This is a rather tight constraint on a country which is developing its

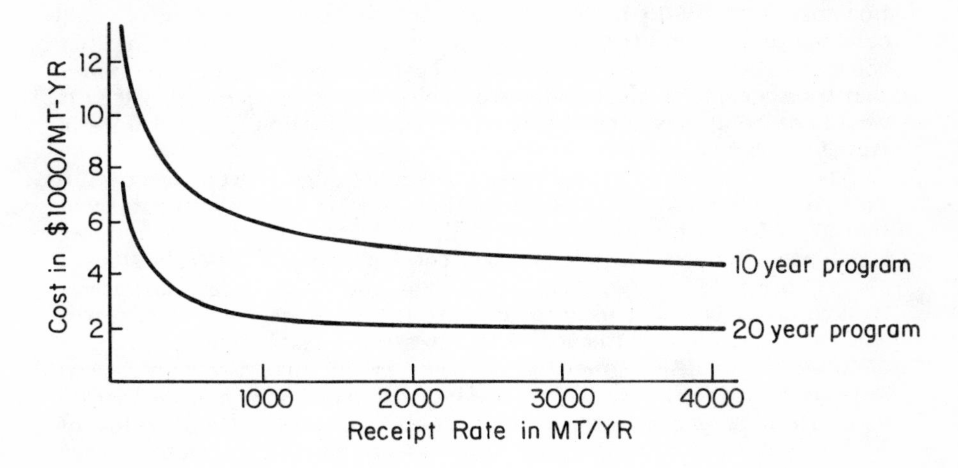

Fig. 8.2. Multilateral storage costs.

spent fuel storage strategy; a shipping cost of $32/kg corresponds to a short interegional distance, such as that between central Europe and The Hague.(7)

However, the breakeven transportation cost for a country with a small nuclear energy program is less of a constraint. A country with an annual spent fuel output of 100MT faces a $20/kg unilateral transporta-

Table 8.3: Breakeven Transportation Costs

10 yr. program

Annual output of spent fuel in MT/yr

	50	100	150	200	250	300	500	1000	1500
10	170	91	63	58	55	44	38	22	17
20	180	101	73	68	65	54	48	32	27
30	190	111	83	78	75	64	58	42	37
40	200	121	93	88	85	74	68	52	47

20 yr. program

	50	100	150	200	250	300	500	1000	1500
10	150	86	63	56	49	44	36	22	17
20	160	96	73	66	59	54	46	32	27
30	170	106	83	76	69	64	56	42	37
40	180	116	93	86	79	74	66	52	47

Unilateral Transportation
cost in $/kg

tion cost and a $96/kg breakeven transportation cost. This country could participate in a multilateral storage arrangement by satisfying this boundary condition on transportation costs. Since the IAEA estimates that transoceanic shipment will cost about $80/kg, economic constraints would not be an obstacle to this country participation in a multilateral storage program.

These multilateral transportation costs are slightly weighted toward a multilateral storage arrangement. For every annual output of spent fuel on Table 8.3, the least-cost multilateral arrangement is a program which has a receipt rate of 5000MT, corresponding to a yearly storage charge of $5333 MT. Therefore, when the least-cost multilateral storage arrangement is used to compute the transportation cost, Cm is a constant. However, it is unrealistic to assume that a large percentage of future regional spent fuel storage regimes will have a receipt rate as large as 5000MT. The breakeven transportation costs for more moderate multilateral examples can be calculated. Consider receipt rates of 500MT and 1000MT, which are estimates for possible regional storage programs in Southeast Asia and South America, respectively. If a 20/kg unilateral transportation cost and a 10 year storage program is assumed for all countries, then the breakeven transportation cost for a country with a 50MT/yr annual output of spent fuel is calculated at $152/kg for a multilateral program with a 500MT receipt rate, and $168/kg for a 1000MT receipt rate. Although both these estimates are less than the corresponding breakeven estimate of $180/kg on Table 3, they are still much higher than the $80/kg IAEA cost estimate for transoceanic shipment of spent fuel.

The numbers in Table 8.3 predict the economic components of policy choices that would be made by the various countries: Countries with large nuclear programs would not ship their spent fuel output over great distances because of breakeven transportation costs. Instead, they would probably have a storage program based within their own borders. The U.S. and the Soviet Union are both in this situation, and both have built their own storage programs. Perhaps political motivations and economic incentives might induce these large countries to base multilateral programs within their own borders.

The controversial question, however, is whether or not the countries with small nuclear programs can be convinced to participate in a multilateral storage program. These are the countries which are reputed to pose a threat to international stability through the horizontal proliferation of nuclear weapons. The goal of an international spent fuel storage regime would be to provide disincentives for these countries so that the acquisition of nuclear-weapons capability does not appear to them as a reasonable policy strategy. Although the economic factor is not the only, or even perhaps the leading, component in an international decision about spent fuel storage, many of the smaller countries weigh economics heavily in their decisions on whether or not to participate in a multilateral storage program; Table 8.3 suggests that they can be convinced, at least on economic grounds, to participate. Otherwise, the breakeven transportation costs which they would face would be higher than the actual cost of transporting spent fuel in a multilateral arrangement.

NOTES

(1) NRC Environmental Impact Statment, March 1978, Project no. M-4, pp. 5-6.

(2) IAEA.

(3) IAEA report, p. 232.

(4) Private communication, Leonard Bennett, IAEA Staff.

(5) IAEA report, pp. 230-233.

(6) There are columns on both cases which exhibit an oscillating behavior that results from the lumpy segments of facility sizes. For instance, country X, which has a spent fuel output of 500MT/yr and elects to use several 400MT facilities in its program, has an abnormally high cost rate of $10,680/MT-yr. This is the result of an assumption that a country uses only one size facility in its storage program. Therefore, this country will have to build two 400MT facilities to accomodate the 5000MT of spent fuel accumulated in a 10-year storage program. At the end of the 10 years, there will still be 3000MT of storage space not utilized in the second facility. Thus, because of high non-utilization costs, this particular situation has a high cost rate relative to the other options open to that country. In the next largest storage program option for countryX, it needs to build only one 5000MT facility. Since the facility is fully utilized throughout the 10 year program, country X faces a much lower cost rate for spent fuel storage. The model predicts correctly that country X will not choose to use two 4000MT facilities in its storage program.

(7) IAEA.

9 Incentives and Disincentives

Daniel Poneman

The essential problem in trying to establish an international spent fuel storage (ISFS) regime is to overcome the inherent divergence of energy interests among states. Finding a worldwide common denominator for any energy policy is difficult; for a nuclear energy policy it is nearly impossible. Different states have different energy resource bases, fossil, hydro, or uranium; they put them to different uses, agricultural, industrial, and domestic; and they pursue different economic objectives.

Outlook, objectives, and practices also differ widely among nations in the pursuit of nuclear nonproliferation. The Soviet bloc countries are compelled to send their spent fuel back to the Soviet Union, but the U.S. cannot exert such compulsion. There have been serious conflicts of nuclear interest among other developed nations; France and the Federal Republic of Germany, for example, are reluctant to sacrifice lucrative reactor export contracts on the altar of nonproliferation. These nations contend that proliferation can best be controlled by selling technology to developing countries under tight safeguards. By refusing to export such technology, they contend, the developed nations will foster resentment in the emerging countries, thus inducing them to develop the technologies independently and without the safeguards which accompany Western exports.

There are also deep divisions within the group of developing nations. Some of the traditionally poor nations have become very rich from oil revenues. The very existence of others – Taiwan and Israel, for example – is threatened. Nuclear energy may appeal to a developing nation as an escape from dependence on imported oil, as a means to advance technological development and thereby reduce the rich-poor gap, or as a symbol of prestige. For a few, it also offers the lure of a quiet nuclear-weapons capability under civilian cover.

Any proposal traversing boundaries of national interest is immediately suspect. For example, American proposals to restrain nuclear technology exports for the sake of nonproliferation are viewed in

Europe as thinly-disguised efforts to regain commercial hegemony in nuclear exports. Similarly, proposals for nuclear restraint are viewed in developing nations as imperialistic. How, then, is a consensus transcending these differences to be established? Since total congruence of self-interest among nations is impossible, ways must be identified to encourage cooperation based on limited common objectives. Furthermore, it may be possible to play various interests and priorities off against one another.

If policy makers can raise the political costs of acquiring nuclear weapons while enhancing the economic advantage of nuclear restraint, they will be on the right road. A spent fuel storage regime could not only remove weapon-source materials from possible nuclear weapon states, but could also bolster the presumption that it is undesirable and antagonistic to develop nuclear weapons. However, these beneficial effects can only be derived from a broad consensus among states that their interests are best served by trading in their short term reprocessing options for other benefits. This paper considers the incentives and disincentives to international cooperation for the storage of spent fuel which might be used to generate such a consensus.

INCENTIVES

In discussing incentives, it is important to remember that one nation's advantage is often another's disadvantage. Attractive economic incentives for the developing nations could beggar the developed countries. On the other hand, proliferation resistance, of major interest to the United States, has little appeal for the developing countries. The following discussion outlines some of the major advantages and disadvantages of spent fuel storage for various interest groups which must be motivated.

Basic Motivations

Averting licensing delays

The central governments in countries with well developed nuclear energy programs could be attracted to spent fuel management by the opportunity to reduce or avert reactor licensing delays. Increasing safety standards, lengthy licensing procedures, and environmental protests have lengthened average reactor lead times in the U.S. from five to over ten years.

In West Germany, opposition to nuclear power is found among anti-industrialists, environmentalists, and segments of the political parties in power, the Social Democrats and the Free Democrats. Party agitation, adverse court decisions, and large demonstrations at reactor sites at Wyhl, Grohnde, and Brokdorf, have moved Chancellor Helmut Schmidt to concede that the disposition of spent nuclear fuel must be resolved before reactor construction can be authorized. The net result has been a moratorium on domestic nuclear power construction in the Federal Republic that may well last three years or more.

In France, Confederation Francaise Democratique du Travail (CFDT), the large trade union, has called for an 18-month moratorium on nuclear construction while the dangers of nuclear energy are fully analyzed.(1) Popular dissatisfaction culminated with the July 1977 demonstration at Creys-Malville, where more than 20,000 demonstrators protested the construction of the Super-Phenix 1300 MWe fast breeder reactor. In a polic barrage of tear gas canisters, one demonstrator was killed and about 100 others were injured.(2)

In Germany, popular concern focuses on radioactive waste disposal, creating a dilemma for nuclear planners: Reactors cannot be built until construction of waste disposal facilities has begun; but if reactor construction slows down enough, there could be growing reluctance to embark upon large-scale commercial waste disposal ventures. Other countries facing this dilemma might be willing to sacrifice reprocessing temporarily in order to break out of this circle, which has the potential to stifle completely their nuclear energy programs. The option to send spent fuel to an international spent fuel storage facility might appear to be the only port in a storm for nations, such as Austria and Denmark, where reactors cannot be built unless waste management arrangements are complete. Similarly, nations facing shortages of spent-fuel storage at their reactors might be happy to part with significant quantities of spent fuel in order to prevent reactor shutdown.

Cost Savings

Two questions must be addressed: Does international spent fuel storage significantly reduce the cost of the nuclear fuel cycle compared to national spent fuel management or reprocessing? And can cost advantages to the states posing a proliferation threat be made sufficient to encourage their participation?

The answer to the first question depends on where the facility is located, how it is built, and how much it would cost to ship spent fuel there (Greer and Dalzell). This cost should be compared to that of reprocessing – less the value of the extracted uranium and plutonium – and fabrication of new fuel elements for either light water recycle or the breeder. Regardless of the absolute cost of international spent fuel storage, the nuclear suppliers could offer subsidies to target states guaranteeing them additional economic benefit for cooperation. Nevertheless, the countries of concern might not find the advantage of free spent fuel storage sufficiently attractive to outweigh limiting their reprocessing flexibility. A nuclear power option, including reprocessing and the breeder, might be worth a great deal in security terms to countries that have experienced quadrupled oil import bills since the 1973 OPEC embargo. OECD forecasts have estimated that the current deficit, including official transfers, of the non-oil-producing developing countries reached $5 billion in 1977.(3) Furthermore, the adverse political reaction could be intolerably severe for the developed countries, given the intense nationalism and sensitivity surrounding the issue of energy independence. Finally, even the appearance of discrimination against the developing countries would immediately jeopardize cooperation. Attempts to influence nations' domestic energy policies through commercial incentives could be viewed as mercenary acts of imperialism, and so provoke developing nations to reject international spent fuel storage as a matter of nationalism and prestige.

Energy Independence

Since energy independence is clearly a central concern of all nations, the best incentive to induce countries to forfeit their spent fuel would be one that directly enhanced that independence. Although spent fuel could theoretically be traded for German coal or North Sea oil, the most likely trade would be for natural or enriched uranium, from countries such as the U.S., South Africa, Canada, or Australia.

Countries may reduce their vulnerability either by minimizing dependence upon foreign energy supplies or reducing its effects. There are two ways to minimize energy dependence upon a single supplier or single group of suppliers.(4) First, a nation can diversify its sources of energy supply. This policy is already being pursued; but it is expensive in the short run because existing patterns of energy imports are usually the least expensive, given the short run inflexibility of energy demand. The second way to minimize dependence on foreign energy suppliers is to increase reliance upon indigenous energy resources. This policy is even more expensive than diversifying suppliers because it requires a shift in both production and consumption patterns. Moreover, the theory of comparative advantage demonstrates that if there is any difference in the cost of producing a given good between two countries, then both can attain higher incomes through specialization and trade.(5) To achieve energy independence, nations must sacrifice this additional income.

Poorer countries may be unable to be selective in their choice of energy suppliers. On the other hand, these countries usually have a rather small commercial energy sector, and so may develop flexible consumption patterns without the great expense of transforming existing social and economic structures.(6)

If a nation chooses not to actually reduce its energy dependence, it can at least reduce the possible harmful effects of dependence by increasing linkages of mutual dependence or by obtaining guarantees that energy imports will not be interrupted. Joseph Nye has noted that country A's economy is "sensitive" if other actors can hurt it, and is "vulnerable" if it cannot successfully react to reduce the injury.(7) Because it is expensive and difficult to reduce dependence, a country might choose to <u>increase</u> linkages, and hence mutual sensitivity, with its trading partners. These linkages could make the partners invulnerable because together they could take retaliatory measures which would render initial unfriendly acts self-defeating.

A country may choose to reduce the risks of dependence by inducing exporters to guarantee that they will supply their commodities under agreed terms. This policy is most effective when there are a number of suppliers competing for shares of the market; competition increases the probability that capricious export policies will drive the customer to another supplier rather than to his knees. Historically, however, commodity agreements have not fared well whenever one party perceived that its best interests were no longer served by the agreement.

What is the relevance of these considerations to spent fuel? The energy value of the uranium and plutonium contained in spent fuel adds to a country's exploitable energy resources, and hence reduces vulnerability to foreign energy suppliers. Under an international spent fuel storage regime, spent fuel could be traded for an assured supply of uranium if there was someone willing to pay for the spent fuel. However, it is unlikely that any country would be willing to pay for spent fuel unless it was given the right to reprocess.

With respect to interdependence, spent fuel is a political hostage for its owner; it is bargaining leverage against those who wish to prevent reprocessing in the owner's territory. The Indian Government seems to be telling the U.S. Government, "If you refuse to send more nuclear fuel to Tarapur, then we will reprocess the uranium and plutonium from the fuel elements you already sent us." Once the U.S. breaks an agreement to send fuel, the importer may feel fewer qualms about breaking his end of the bargain.

Policy options

What policies can be adopted to induce nations to surrender their spent fuel to an international spent fuel storage regime? First, diplomatic pressure can be used. The Republic of Korea refrained from buying a reprocessing plant from the French under American pressure. This concession may have been motivated by the fear of reduced U.S. support for the Park regime, or by the advantages to be gained from a political IOU from the Americans. In any event, the Koreans can always threaten to proceed with reprocessing plans at a later date if they become dissatisfied with American policies.

American diplomatic pressure has been less successful in Pakistan and Brazil; neither nation is as dependent upon U.S. security guarantees as is the Republic of Korea. Former Prime Minister Bhutto and President Geisel were concerned not only with Indian and Argentine nuclear capabilities, but also – perhaps more important – with sustaining domestic popular support; a firm, unyielding response to the U.S. would be much more popular than giving in to American pressure.

Second, an international spent fuel storage option could offer countries the opportunity to avert the political and environmental costs of domestic disposition of spent fuel. The chance to escape the domestic political problem of waste management might attract some of the developed nuclear-power countries, such as West Germany, Austria, and Denmark. However, the countries most likely to "go nuclear" in the next 20 to 30 years are often unmoved by environmental concerns. In the Stockholm Conference on the Human Environment, developing countries cited economic growth as a much higher priority than pollution control.(8) Indeed, concern for the environment was characterized by some participants as a capitalist ruse aimed at perpetuating the subjugation of the Third World.

Third, an international spent fuel storage regime would offer a country the opportunity to convince its neighbors to send their spent fuel to an international facility rather than to reprocess it domestically. Not only would this enhance the nation's own security, but it would save the expense of matching neighbors' reprocessing ventures. This motivation led many countries to ratify the Treaty of Tlatelolco and the Nuclear Nonproliferation Treaty.

Fourth, countries could be offered direct financial or fuel supply compensation for sending their spent fuel to an international spent fuel storage facility. Perhaps the most effective incentive would be to exchange supplies of natural or low-enriched uranium for the fuel value of spent fuel.

There is a trade-off, however, between reducing incentives to develop national reprocessing and breeder facilities and maintaining international control over the disposition of dangerous nuclear materials. Although perfect fuel assurances reduce the incentives for a recipient nation to acquire sensitive facilities, they also reduce the exporter nation's leverage to prevent proliferation through the control of nuclear exports. Given the proclivities of such nations as Pakistan, Taiwan, and the Republic of Korea, such a policy, utterly dependent on the good will of the importer, could be very dangerous. In addition, enriched uranium trade is so politically sensitive that conditions are always attached to its sale. Australia and Canada have demonstrated that political concerns may influence natural uranium trade as well. There will probably be no free market in nuclear fuels in the foreseeable future. Thus, a total, nondiscriminatory fuel inducement regime is, at least in the short-term, both infeasible and undesirable.

There is a fundamental paradox in the way fuel assurances relate to nonproliferation policy: Suppliers dissuade importers from building their own facilities through fuel assurances, thereby guaranteeing the dependence of the importers; at the same time, these importers are urged to accept fuel assurances as a means of reducing their energy dependence. The more insistent the nuclear suppliers appear, the more such assurances are seen by the importers as ensuring dependence rather than independence. Fuel assurances, which clearly distinguish between suppliers and recipients, will not be effective in discouraging proliferation. Because they highlight the dependence of some nations upon others, they cannot continue to deter construction of national fuel facilities in target countries. This does not mean that fuel assurances should not be used at all; they offer, at best, a breathing spell during which more permanent nonproliferation policies can be institutionalized.

Fuel availability

Before placing too much reliance on the notion of fuel assurances, it would be wise to determine whether or not there is sufficient natural uranium and enrichment capacity to make this option attractive. Nuclear News reported that, as of June 30, 1977, the 63 nuclear power plants* in commercial operation in the U.S. generated a total of 45,398 MWE. At the same time, 188 reactors were in operation, under construction, or on order outside the U.S. If all of the nuclear power plants now under construction or on order are completed, the total world generating capacity will reach 391,871 MWe, of which 191,885 MWe would be outside the U.S. Extending projected installed nuclear capacity to the year 2000, the U.S. Energy Research and Development Administration (ERDA) presented a low case for foreign nations, excluding those with centrally-planned economies, of 485,000 MWe.(9) Recent official estimates forecast U.S. capacity at that time to be about 380,000 MWe.(10)

With a once-through uranium fuel cycle, a 1,000 MWe light water reactor will require 5,250 short tons of natural uranium (U_3O_8), or 4,773 metric tons (MT) U_3O_8 assuming a 0.2 percent tails assay, 30 year reactor life, and 65 percent average capacity factor.(11) At this level, total world U_3O_8 needs for reactors on line, under construction, and ordered would be approximately 1,870,400 MT. If we combine the ERDA low foreign case and the 380,000 MWe assumption for American on line capacity by 2000, total capacity of 865,000 MWe will require 4,128,645 MTU_3O_8.

Are there sufficient uranium resources at a reasonable price to accommodate this level of nuclear power capacity? The Congressional Office of Technology Assessment has estimated that, at a cost range of up to $30 per pound, there are reasonably-assured resources of 2,041,000 MTU_3O_8 worldwide, and 1,873,200 MTU_3O_8 estimated additional resources for a total of 3,914,200 MTU_3O_8.(12) If the price were to rise, more resources would become available to accommodate increased demand. If all estimated additional resources are, in fact, real uranium deposits, there will be nearly enough uranium to fulfill projected demand. Furthermore, if nuclear capacity projections were to fall from their present unrealistic heights, reliance on the estimated additional resources would also diminish.

With a total world generating capacity of 835,000 MWe, separative work requirements – assuming a 0.25 percent tails assay, 3.0 percent U-235 – will reach 50.7 million separative work units (MSWU) annually by 1990, and 83.5 MSWU by 2000.** According to Westinghouse, total enrichment capacity will reach approximately 70 MSWU annually by 1990. Thus, there will be excess enrichment capacity in the years to come; and there is time to decide how to proceed with the next increment of enrichment, even assuming a 10 year lead time for enrichment plants. The excess separative work capacity could be used to drive down the tails assay, thereby reducing uranium feed requirements.

In sum, under almost any conceivable set of circumstances, there

*Including all reactors over 30 MWe in size.

**Separative work units (SWU) are used to measure enrichment capacity.

will be sufficient enrichment capacity to satisfy the demand for assured fuel supplies. Whether there will be enough uranium ore depends on the price, the rate of nuclear power growth, and the reliability of estimates of additional uranium resources; current trends indicate that it is reasonable to offer fuel assurances.

If it can be concluded that aggregate world uranium resources and enrichment capacities are probably sufficient to meet requirements, the next problem is how to distribute those resources and services and to allocate the revenues they accrue. Fuel supplies or services are typically offered by nuclear reactor vendors to make their package more attractive. The real competitive drive, and the profits derive primarily from the nuclear power plant itself. If fuel services were internationalized, an alternative method for reactor competition would have to be devised. The problem would be to decide along what lines to divide the market in order to keep all nuclear reactor suppliers happy without compromising nonproliferation objectives.(13)

Perhaps the only way to satisfy importer desires for secure and diversified supplies and to avoid windfall profits and the inefficiencies of investment allocation, would be to establish an international nuclear fuel bank under the aegis of the International Atomic Energy Agency. Suppliers and importers would receive shares in the pool in return for natural or enriched uranium and spent fuel, respectively. Spent fuel could be stored at an international spent fuel storage facility. With the credits obtained for the energy equivalent of their spent fuel, importers could purchase nuclear fuel, while suppliers would receive proceeds from the sales in proportion to their contributions to the pool.

There are some problems with this approach. First, to the extent that there is excess enrichment capacity, as well as existing enrichment contracts unevenly divided among the suppliers, the disposition of different enrichment facilities would be subject to dispute. Some could argue that the proposed Coredif, for example, was superfluous, and that its owners should not be rewarded for overexpansion. The French could reply that the surplus separative work capacity would be used to conserve natural uranium by using it to strip more U-235 from each kilogram of natural uranium feed.

Second, nuclear power is a commercial venture, and the requisite convergence of interests or desire for cooperation among suppliers might be lacking as long as the U.S. controls a disproportionate share of the market.

Despite these practical objections, fuel assurances in the context of an international fuel bank could be arranged under certain circumstances. If the domestic nuclear programs of key suppliers continue to be stymied, an international fuel bank might be worth some commercial loss to ensure their survival. The detonation of another nuclear device or aggressive militarism by non-nuclear-weapon states, the generation of a new concensus on internationalization of the fuel cycle in the INFCE, or a continuation of the trend of nuclear suppliers to take their responsibility vis-a-vis proliferation more seriously are all factors which could make fuel assurances appear politically and economically advantageous.

Negative Aspects of Fuel Assurances

Before we accept fuel assurances as an unmitigated good, it is necessary to examine their impact upon the status quo with relation to nonproliferation objectives. The reaction to dismal domestic nuclear prospects in France and West Germany has been to push heavily into the export market, both to accrue valuable export earnings in the short run and to sustain the nuclear industry long enough to weather the domestic anti-nuclear storm. These efforts may fail. First, if the effect of anti-nuclear demonstrations is as important as the Germans contend, then the increasing failure of domestic nuclear programs will discourage exports. Many developing countries want nuclear energy in order to ride the crest of the technological revolution; if nuclear energy stalls in the developed countries, the emerging countries will not have to worry about being left behind. They may well be happy to forego the massive loans and expenditures of foreign reserves needed for nuclear reactors.

Second, nuclear energy may be inappropriate for the energy needs of many developing countries. For those lacking in uranium or technical resources, the substitution of nuclear dependence for oil dependence would not be a great boon. One should consider that nuclear power plants became popular as a substitute for large coal-fired plants in large electrical grids. If a single generating unit provides over 15 percent of the electricity to a grid, its failure can easily trigger the collapse of the whole grid.(14) Most developing countries lack sufficiently large grids.(15) For example, Westinghouse is building a 600 MWe reactor in the Phillipines for a 2300 MWe grid; the remaining 1700 MW of capacity might not be able to handle the load at the time of failure.

Under present conditions, an era of limited light water reactors and very limited breeder reactors could be forthcoming, gradually yielding to new energy sources in the next century.

Fuel assurances might provide the needed lubricant to get nuclear power back on the track. France and West Germany, relieved of the immediacy of their waste disposal dilemmas, could proceed down the breeder road untrammeled. Export offers would become more attractive as new nuclear power plants began operating in the developed countries, and as rising export volumes and increased standardization led to reduced reactor price tags.

Also, this course of events would reinforce the presumption that nuclear power is a key to future development rather than an albatross around the necks of the developed countries. This perception, coupled with blandishments and incentives to cooperate in international spent fuel storage, might firmly entrench nuclear energy programs where there otherwise would have been little commitment. However, an international regime that expanded nuclear programs to countries which would not otherwise have had them would lay the groundwork for an increase, not a decrease, in proliferation potential.

DISINCENTIVES

Transport and legal arrangements, management and operational control, and safeguards present hurdles that must be overcome before international cooperation for spent fuel storage can be realized. Most of these problems are discussed in other chapters. Loss of autonomy and expense merit an added word here.

Loss of Autonomy

At present, many nations do not wish to part with all of their spent fuel. Reprocessing offers the possibility of an independent breeder program, and will not easily be forsaken.

This disincentive could be minimized if an international fuel bank could attain widespread acceptance. Countries would have less need to worry that others might "get the jump" on them in reprocessing. Also, a country would be able to use its credits to obtain oxide fuel if an international reprocessing and oxide fuel fabrication center were eventually built.

Expense

If it does not reprocess, a developing country could give away waste and get good fuel in return. Undeniably, however, the costs of transporting spent fuel to the international or regional facility would have to be covered.

The cost of uranium mining, milling, conversion, and enrichment might be excessive for developing countries, and especially for those nations already in financial straits. Countries such as Peru and Turkey are already risking default on foreign loans; if this happens, credit will become even more restricted for developing countries.

The nuclear industry as a whole has been losing money.(16) Without governmental support, an international spent fuel storage regime may not be commercially viable. Giving financial aid to the nuclear industry solely for the sake of preventing proliferation is like trying to put out a fire with kerosene. Even governmental support is limited; the Eximbank will not continue indefinitely to invest in nuclear power if short term losses are not recovered.

SUMMARY

A successful international spent fuel storage regime must offer something to nearly everyone. Developing countries might be attracted by fuel assurances; developed countries, by a partial solution to their domestic nuclear problems; and some countries, by the desire to stem nuclear proliferation through international cooperation. It is debatable whether there are sufficient natural uranium resources to meet future needs. How long will they last, and what price will they command?

Finally, before any major initiative is undertaken, all ramifications must be fully examined. Would fuel assurances encourage or retard proliferation may be a question whose answer is antithetical to the one originally projected.

Political considerations seem likely to continue dominating spent fuel policy in developed and near-nuclear-weapons states. The desire to reduce energy dependence by building up nuclear energy programs is great, but reprocessing and recycle appear to have little impact upon the cost of nuclear power. Disguising an international spent fuel storage policy as a commercial transaction may involve too many nations at too high a cost. Also, such a policy would not even touch the near-nuclear-weapons states most fundamental concerns; they want the use of spent fuel to express energy independence, to exert leverage upon the developed nations, and to otherwise enhance their international clout. The economic approach is clumsy; to think that it could cajole nations into surrendering their spent fuel would be over-optimism. It would probably be wise to attack an essentially political problem in a political manner.

NOTES

(1) Charles Hargrove, The Times (London), Oct. 21, 1977, p. 8.

(2) William Sweet, "The Opposition to Nuclear Power in Europe," Bulletin of the Atomic Scientists, Dec. 1977, pp. 41-47.

(3) OECD, Economic Outlook, Paris, No. 20, December 1976, Table 24, p. 67, cited in Guy de Carmoy, Energy for Europe: Economic and Political Implications, (Washington: American Enterprise Institute for Public Policy, Research, 1977), p. 23.

(4) See Mason Willrich, Energy and World Politics (New York: The Free Press, 1975).

(5) See Richard E. Caves and Ronald W. Jones, World Trade & Payments: An Introduction, 2d ed. (Boston: Little Brown & Company, 1976), Chapter 2.

(6) France possesses around 55,000 metric tons of natural uranium reserves (at up to $30/lb.) and has access to an equal amount in its former African colonies.

(7) Nye, J.S., "Independence and Interdependence," Foreign Policy, Spring 1976.

(8) "What Happened at Stockholm," Bulletin of the Atomic Scientists, September 1972, pp. 16-56. Of course, pollution control sometimes promotes economic growth.

(9) Congressional Research Service, *Nuclear Proliferation Factbook* (Washington: General Printing Office, 1977), p. 225.

(10) Unofficial estimates are even lower, dipping well below the 300 reactor mark.

(11) Thomas B. Cochran, Russell E. Train, Frank von Hippel, Robert H. Williams, *Proliferation Resistant Nuclear Power Technologies: Preferred Alternatives to the Plutonium Breeder*, (to the ERDA & MFBR Review Steering Committee, April 6, 1977), p. III-3.

(12) *Nuclear Engineering International*, November 1976, cited in U.S. Congress, Office of Technology Assessment (New York: Praeger Publishers, 1977), p. 249.

(13) For a fuller treatment of the issue of nuclear market sharing, see Abraham Ribicoff, "A Market-Sharing Approach for the Nuclear Sales Problem," *Foreign Affairs*, July, 1976.

(14) Denis Hayes, "Nuclear Power: The Fifth Horseman," Worldwatch Paper 6, May 1976.

(15) Richard J. Barber Associates, Inc., *LDC Nuclear Power Prospects, 1975-1990: Commercial, Economic, & Security Implications* (Washington: U.S. Energy Research and Development Administration, 1975), pp. II-8-10.

(16) See, for example, William Mety, "Nuclear Goes Broke," *The New Republic*, February 25, 1978, pp. 23-25; and Richard Masters, "Standardization and Quality are Priorities at Kraftwerk Union," *Nuclear Engineering International*, August 1976, p. 43.

IV

The Broader Political Context

10 Public Response to Nuclear Energy

Dorothy Zinberg

INTRODUCTION

The success or failure of new policy initiatives in the management of the nuclear fuel cycle, and especially of the part known as the "back end," is likely to depend on the domestic political acceptability of those measures in the western democracies involved. Accordingly, assessing public reactions in the major nuclear energy countries will comprise an essential part of the analysis of any new proposal such as international cooperation in the storage of spent nuclear fuel.

Spent fuel management will be a particularly troublesome political issue, in part because of the difficulty in distinguishing between spent fuel and nuclear waste. If reprocessing of nuclear fuel is not merely delayed but delayed indefinitely or avoided altogether, spent fuel storage is likely, in the public mind, to become tantamount to permanent nuclear waste disposal and, therefore, fraught with all the political difficulties surrounding that ultimately necessary activity.

John Deutch, Director of Energy Research for the Department of Energy, in recent testimony(1) identified the importance of assuaging "legitimate public concerns" through the development of a "responsible nuclear waste management program."

How the public perceives the ways in which nuclear wastes can be safely dealt with, and how these perceptions are shaped by the scientific technological community bear careful scrutiny.

THE RISE OF PUBLIC PARTICIPATION IN NUCLEAR DECISION-MAKING

Part of the complexity of policy-making for nuclear energy stems from the increase of public participation in government decision-making from decade to decade following World War II. As seemingly disparate, controversial issues surfaced, so did the public interest. Concerns about fluoridation, new drugs (particularly thalidomide), DDT, airport sitings, highway development, automobile pollution, industry-related carcinogens, recombinant DNA research, and, of course, nuclear power plants were exacerbated by a growing preoccupation with the fragility and finiteness of the planet.

Rachel Carson's Silent Spring(2) received increasing attention during the decade following its publication; by 1972, Limits to Growth(3) triggered even more intense debates. That the response was not simply American can be judged by its sales record abroad; more than one-quarter of a million copies were sold in the Netherlands within the first month of its publication.

A shrinking economy and a disenchantment with some of the social costs of technology provided the background from which citizens groups emerged, while the Freedom of Information Act in July 1967 provided access to heretofore secret documents. Social responsibility, accountability, regulatory procedures, and environmental impact statements became basic words in a new vocabulary.

Nuclear power plants did not begin to attract public attention until 1969; a group in Minnesota challenged the AEC and the utility company over the running releases of radioactivity from nuclear power stations and fuel reprocessing plants. Although they lost their case, they succeeded in drawing national attention to their cause. The Minnesota Committee for Environmental Information was formed from this group.

In 1970, the Environmentalists, a loose coalition of related public interest groups and individuals, gained national prominence through their Earth Day activities.(4) By 1971, in addition to grass roots activities, the Environmentalists had added a Washington base, the Consolidated National Interveners, whose target was the U.S. Atomic Energy Commission's safety inadequacies. The Environmentalists had found a unifying theme – nuclear power.

Through the efforts of Ralph Nader's Citizen Action Group, members of organizations such as Friends of the Earth, the Union of Concerned Scientists, the Committee for Nuclear Responsibility, and the National Interveners organized a conference, Critical Mass '74 which aimed at establishing a nuclear moratorium.(5) By 1975, Critical Mass had become a national organization. It and similar interest groups introduced a number of public initiatives for a moratorium on nuclear energy to the voters in Arizona, California, Colorado, Montana, Ohio, Oregon, and Washington. Although they were defeated by a margin of approximately two to one, this did not dim the enthusiasm of the antinuclear forces who are preparing initiatives in another nineteen states. Other states have already passed referenda stating that they will not reprocess any nuclear waste, provide storage facilities, or allow nuclear wastes to be transported across state boundaries.(6)

SHIFTING PUBLIC CONCERNS

To sort out what is rational from what is irrational in public attitudes toward nuclear power is a major challenge after thirty years of intertwining psychological and realistic fears. That some progress has been made can be deduced from the absence of, or at least a reduction in, newspaper accounts and public opinion polls that link nuclear power plants to the earliest nuclear image – the mushroom cloud. A decade ago much of the public feared that nuclear power plants could explode like nuclear weapons. Such views have clearly diminished. Perhaps the change was brought about by the fact that reactors have been in operation since 1947, and, despite a number of accidents, there has never been an explosion to validate the fears.

However, other preoccupations have taken precedence. Fears of the effect of natural catastrophes such as earthquakes on nuclear reactors, seepage of radioactive material into cooling water, accidental disruption of the cooling water supply leading to core melt downs, alteration of sea water temperature, circulation of enormous volumes of sea water through cooling systems and the concomitant destruction of marine larvae, design failure, or human failure have gained increased national attention. More recently, two related issues have moved center stage: One is the increased potential for the diversion and theft of enriched uranium and plutonium; the other is the problem of dealing with nuclear wastes. Because the former is dependent to some extent on the successful resolution of the latter, it is the latter which has in the past two years become the primary focus of many scientists, federal agencies, public interest groups, and the courts.

CENTRALITY OF WASTE DISPOSAL ISSUE

According to Washington wisdom, "A budget is a policy statement." In 1978, the federal budget allotted $126 million to commercial waste management programs.(7) Only eight years before, this figure was zero; in 1976, it leaped to $10 million; and it is projected to be $163 million in 1979. The military waste management budget, which was $20 million in 1970, has also escalated, although not as rapidly as the commercial budget; it will reach $282 million in 1979. Thus the combined research and development funding in this area will be $445 million in 1979.

These figures reinforce the perceptions derived from public opinion polls, newspaper coverage, scientific meetings, and publications during the past two years; the budget does indeed reflect the centrality of waste disposal as the critical issue in the future growth or decline of nuclear power.

A British survey(8) of public opinion on nuclear power stations commissioned by New Society revealed that the overall response was the very characteristic one of "stoic acceptance," but it also found that one out of two people considers the long-term disposal of radioactive waste at sea or underground an unacceptable risk. The poll was conducted in late 1974 before the secrecy of British policy making was significantly penetrated by the Flowers Report and, even more vividly, by the Windscale investigations. Opponents of the proposed expansion of the Windscale nuclear fuel reprocessing plant have charged that, by storing waste from other countries, spent fuel, Britain would in effect become the "nuclear dustbin of the world."

Because Great Britain has a much longer history of tight secrecy in government than the U.S., the public in Britain has been unaware of almost all nuclear-related decisions. The past year has seen a quickening of public interest in waste storage issues, brought on by the Windscale furore. "There is almost unanimous dissatisfaction with existing and planned methods of waste storage." Even some pro-nuclear groups have pronounced them "ethically and technically intolerable."(9)

The case has yet to be decided but the British public is focused on the same point as the Americans interviewed in an intensive Harris survey: "...on direct questioning, the public views radioactive waste disposal as the biggest problem connected with nuclear plants."(10)

An OECD Report made the following statement: The (nuclear safety) debate focuses on the problem of radioactive waste disposal which is central to the reservations about nuclear programmes to be found in most developed countries, despite the fact that nuclear power is also regarded as essential for the supply of energy.(11)

Harvey Brooks has predicted that "...should nuclear energy ultimately prove to be socially unacceptable, it will be primarily because of the public's perception of the waste-disposal problem."(12)

And John Deutch informed the House Science and Technology Committee that "the importance of showing the nation that we can take care of spent fuel is paramount at this point."(13)

It is clear that consensus obtains as to the problem; how, or if, it can be resolved as the public becomes more active, informed and, desirous of participating in decision-making is less so.

PRESENT TRENDS

At a time when action on waste disposal and other fuel cycle problems related to proliferation are gaining momentum, the strength of public opinion against nuclear power appears to be growing. In 1975, the Directory of Nuclear Activists recorded 149 anti-nuclear organizations. Harris polls, which in other years had reported that approximately 60 percent of the public was in favor of nuclear energy and 30 percent against, detected an anti-nuclear shift in 1978, with figures of 48 percent and 36 percent respectively.(14) A nuclear accident or another oil embargo could produce an entirely new set of figures, but overall the anti-nuclear forces appear to be gaining.

A recent study came to the following conclusions:

> Our immediate prognosis is for extension rather than diminution of the opposition to nuclear technology. Public opinion, which has consistently supported nuclear power, is nonetheless deeply divided, much as it was during the war in Vietnam. There is some evidence that wider public exposure to rancorous debate on nuclear power may well stiffen the opposition...(15)

Although anti-nuclear efforts started somewhat later in Europe, they gained rapid momentum. For example, citizens of Wyhl, Germany, a grape-growing community, became concerned that climate changes would result from a nuclear installation; they have managed to block construction of the proposed power plant. And at the European Community headquarters in Brussels, the majority of people interviewed in the Energy Research Development Sector in 1977 expressed the belief, and certainly the hope, that anti-nuclear activity had peaked.

However, the head of Power Plant Technology took quite another view and expressed his feelings very strongly:

> I don't believe in the reaction of societies. They are manipulated by lobbies. Slowing down progress is a very anti-democratic process. Parliaments and politicians have been elected to represent the society. I don't think public opinion exists; it is mentioned by self-serving interests. One should not confuse the problems with the reactions. The majority has its rights, and it is pro-nuclear.

The Director of the Division for Energy Research forecast that "Europe as a whole will go nuclear in a way. It will overcome the anti-nuclear movement. The public will tire of anti-nuclear movements. The trade unions in Germany are becoming pro-nuclear because they are afraid they will lose their jobs."

Yet other Europeans disagreed. In Germany, the International Institute for the Environment and Society, which has completed a study on nuclear power plant siting in West Germany(16) believes that public participation has not diminished. Environmental action has led to political action which has substantially slowed down the pace of nuclear power development.(17) Those nuclear power plants now operating or on line will be made more secure with new safety devices, and no new plants will be built until safe waste disposal facilities are found. Defining "acceptable safety" has evolved into prolonged legal battles. Thus, in effect, anti-nuclear public groups have brought about a moratorium on nuclear-power-plant building in Germany.

William Sweet, an American historian who spent 1976-77 studying the opposition to nuclear power in Europe, writes:

> What I failed to appreciate, when I set off for Europe, was the degree to which people responsible for nuclear matters were seeing their freedom of action limited by anti-nuclear forces...groups opposing nuclear energy have registered startling gains in Sweden, Great Britain, West Germany and France, establishing themselves as distinct political forces. They embrace people of all ages and occupations, but educated middle class people predominate, and the most influential leaders tend to be in their early thirties. While the opposition to nuclear energy is in large part rooted in the youth movements of the late 1960s, these movements have evolved in unique ways in Europe.(18)

Government officials in France express much the same opinions as officials in Germany and the European Community. They hope the peak of public protest has past." Unions concerned with unemployment and a slow-growth economy are reversing their anti-nuclear attitudes, although the fact that the Confederation Francaise Democratique du Travail (CFDT), the second largest union in France, has asked for an 18-month moratorium on nuclear power plant construction seems to indicate there is somewhat less agreement about protest movements than officials would like to admit.

Ecologists have formed political parties in several countries. In France, the ecologists gained 9.3 percent of the national vote in municipal elections in March 1977; and, in a recent French public opinion poll, 56 percent of those questioned were able to define ecology correctly. What is even more striking is the fact that "77 percent believe that ecology should have as important a place in educational curricula as history and geography."(19)

Although French officials insist that the July 1977 confrontation at the construction site of the Super Phenix fast-breeder in Creys-Malville was a non-recurring, unique event, police killed a demonstrator and gave the opposing forces their first martyr.(20) Perhaps the officials' prediction will prove accurate, but at present the confidence appears to be based on wish – a less than substantial foundation on which to keep an industry alive.

The Swedish government has attempted to organize growing public interest in nuclear energy controversies by setting up 7,000 discussion groups throughout the country, with the hope that open forums will diminish the polarization evident elsewhere.(21) To date the results suggest that the more information the public has obtained, the more convinced the anti-nuclear members have become of their original position. Yet this experiment in participatory democracy might hold the key for similar experiments in countries where feelings have become polarized. As in Germany, new nuclear power plant construction in Sweden is at a temporary halt.

In Belgium, where nuclear energy has had a long and untroubled history, a nuclear plant accident on January 13, 1978 was made public by the Environmental Protection Society. The claim made at the opening of the European Commission's nuclear energy discussions was that 80 people were contaminated by radium-131 and that government authorities had covered up the affair.(22) The Friends of the Earth argued that the major issue was one of disclosure.

Each of the countries mentioned so far – the U.S., Germany, France, Great Britain, Sweden, and Belgium – differ in customs, political institutions, and legal systems. The extent to which the public has been involved in decision-making and the level of secrecy accepted as a given in each of these countries has also varied; but what has become increasingly apparent, and dismayingly so to pro-nuclear governments, is that the past is no guide to the present or predictor of the future.

Furthermore, protestors are not limited by national boundaries; they can migrate quickly from one reactor site to another in Western Europe. Friends of the Earth can be counted on to send a representative anywhere. In America, anti-nuclear groups can muster support from California to New Hampshire. The protestors are now an international brigade.

SECRECY --THE LEGACY OF WORLD WAR II

It would be overly simplistic to claim that public participation in nuclear power issues arose only in response to government and bureaucratic secrecy, or that the debate would have been resolved or at least been less rancorous if the "sunshine laws" had existed since 1945. What is apparent, however, is that citizen participation groups are behaving as if secrecy is a major issue and are demanding greater access to reports from private industry and government bureaucracies. Protection of the environment is the rallying cry of most citizens groups, one of the coalescing factors is the shared belief that the public is not being informed of the real risks involved in the use of nuclear energy; this translates into more of secrecy.

That the public response, whether positive or negative, was not considered in the early days of nuclear energy is not surprising when nuclear power is viewed in its historical context. Many of the senior scientists who today are actively involed in nuclear energy came of age professionally during World War II when they were drafted or volunteered for the Manhattan Project. As their ever-proliferating autobiographical accounts reveal, no small part of the tension and excitement surrounding the enterprise sprang from the secrecy in which it was shrouded. Secrecy was the hallmark of the Manhattan Project and almost every other project that touched even tangentially on the war effort. It was not in the secrecy of the university or the industrial laboratory, but in total blackout secrecy that the first nuclear weapons were developed.(23)

It is difficult to recapture the intensity of the commitment to a national effort that made it possible for normally hyper-curious people to move to a remote part of the country and radically alter their personal and professional lives. Many of the scientists could only guess at what they were developing because they had only partial pieces of the puzzle. At Los Alamos, the activities and correspondence of the scientists and their families were sharply curtailed.

The public, unaware of this particular weapons development, was equally committed to the notion of secrecy. Posters were ubiquitous: "A slip of the lip can sink a ship"; "Careless talk costs lives"; and the British warned, "Tittle-tattle lost the battle." Millions grew up socially and professionally to accept secrecy as a given; it became a way of life and a habit.

The end of the war did not bring about a sudden reaction to previous restrictions; on the contrary. The Cold War, not totally unrelated to United States secrecy restrictions concerning nuclear weapons, followed closely; and the McMahon Act of 1946, a product of Cold War fears, effectively cut off international exchange of nuclear-related information.

The destruction of Hiroshima and Nagasaki relaxed the secrecy ground rules, if only marginally. In the open debate that followed the bombings, the public learned that there had been feverish efforts on the part of a group of scientists who had developed the bomb to stop its use. The Franck Report, written in June 1945 and signed by seven leaders of atomic bomb development, had attempted to warn the government of the awesome consequences which could be expected were the bomb to be used; it foretold a nuclear arms race. There is some doubt that the Secretary of War, Henry L. Stimson, to whom the report was addressed, ever paid it much attention. The Szilard Report, which followed in July 1945, was signed by 88 scientists; others, who were prepared to sign, were stopped by the military lest the state of the progress of the bomb be revealed. It never reached its destination, President Harry S. Truman. Once more, secrecy triumphed.

This secrecy, which served the U.S. so well during the war years, became dysfunctional in the post-war world. Many operations of the Atomic Energy Commission indicated a marked hostility toward the American public. The misdeeds of the AEC's bureaucrats have been well documented.(24,25,26) Suffice it to say here: they lied. Miscalculations about radioactive debris and the ways in which fallout was carried by the wind from above-ground weapons testing in Nevada were carefully kept from the public. The public was exposed to low-level doses of radiation throughout the 1950s; nursing mothers were found to have iodine-131 in their milk; and babies had strontium-90 impregnated in their bones and teeth. These are facts that were disclosed despite the AEC, not because of it. Added to this were the horrors of the Bikini atoll tests, where Japanese fishermen were showered with radioactive ash and once fertile atolls were laid waste and left uninhabitable. This series of events marked the first major breach in the public trust. The images from these episodes – the mushroom cloud, the unseen radiation that seeps into bones, the radioactive-ash-covered fishermen of the Fukuryu Maru – have remained to haunt the proponents of nuclear power. Although weapons testing went underground in 1963, public distrust surfaced.

Thus, for many, nuclear energy is a force that had its origins in war, destruction, deceit, and secrecy. These negative antecedents cannot be completely detached from nuclear power's peaceful applications. Nuclear weapons production continues apace, and the nuclear wastes generated from military weapons programs constitute some 90 percent of the total nuclear waste awaiting storage and disposal. Industrial wastes have been managed far better than those of the military. The linkage to war and secrecy persists.

FROM SECRECY TO SKEPTICISM

As the layers of past secret activities have been peeled back, the public has become increasingly skeptical. The rapid progression of the Vietnam War and Watergate revelations hardened the skepticism, while once sacred institutions and professions began to drop in public esteem. This has not been solely an American phenomenon. While Louis Harris found that 18 percent of his respondents had trust in the press and eight percent in politicians, the New Society study found that journalists and politicians were trusted by only four and five percent of the public, respectively. Similar studies conducted in the U.S. over the past 15 years by Yankelovitch have shown a steadily rising curve of distrust and skepticism of formal institutions.(27)

Scientists, however, have maintained a more consistent position of public trust and can be expected to play a more influential role in decisions relating to ncuelar problems. In the New Society study, which found public trust of the media and politicians so low, scientists were trusted by 68 percent of the population. A comprehensive Harris survey of the public attitude toward leadership in the U.S. found a similar degree of confidence in scientists. Table 1 reveals the differences among specific groups and their attitudes toward scientists.

Environmentalists may look to environmentalists for the final word on nuclear energy, but the public looks not to environmentalists, not to government leaders, and not to the media: rather it looks to scientists. Harris points out that, in fact, "Scientists inspire confidence in people

Table 10.1

"A Great Deal" of Confidence in What Various People or Groups Say on Matters Concerning Nuclear Energy Development

	Total Public %	Political Leaders %	Business Leaders %	Regulators %	Environmentalists %
Scientists	58	63	71	62	37
Nuclear Regulatory Commission (formerly Atomic Energy Commission)	38	31	49	38	6
*Energy Research and Development Administration	35	37	39	32	6
Leading environmentalists	26	22	8	13	57
The President of the United States	24	20	39	24	8
News commentators on television	22	2	2	6	8
News commentators in newspapers	18	8	2	6	8
Heads of electric power companies	18	12	39	7	2
Foundation reports	17	36	14	37	27
State governors	9	12	–	4	6
U.S. Senators	8	2	–	–	4
U.S. Congressmen	7	2	–	–	2
Ads in newspapers by those who support nuclear power	7	–	4	2	–
Labor union leaders	7	–	–	2	8
Adds in newspapers by those who oppose nuclear power	6	–	–	–	14

*Error: ERDA also formerly part of AEC

on both sides of the fence: those who favor more nuclear plants in the United States (64 percent have 'a great deal of confidence') and those who are opposed (53 percent)."(28) The public and the different leadership groups, with the exception of the environmentalists, agree that it is only the scientists who can instill confidence in the future development of nuclear energy.

DISSENSION AMONG SCIENTIFIC EXPERTS

> ....The public perception and acceptance of nuclear energy appears to be the question we missed rather badly in the very early days. This issue has emerged as the most critical question concerning the future of nuclear energy.(29)

Alvin Weinberg is the doyen of nuclear fission development; the "we" he refers to are the scientists and engineers who were pioneers in the nuclear energy industry.

Most ostensible dichotomies in the energy arena are not really dichotomies, but rather overlapping categories with vague boundaries. For example, scientists are not only members of the scientific community; they are also part of the public. And the public is composed of so many disparate elements that any reference to "the public", if not elaborately qualified by age, sex, income, education, ethnicity, geography, occupation, voting behavior, and influence, tends to obscure rather than clarify.

Scientists who had worked to end the war, but who had opposed the bomb's deployment, went public after 1945. The attitude expressed in Oppenheimer's doleful confession, "Scientists have known sin", and the knowledge that the world would never again be non-nuclear forged new coalitions among previously apolitical scientists.

That there was a belief among others that scientists should not interfere can be seen – somewhat ironically in historic perspective – in the exchange between Teller and Oppenheimer. When Teller consulted Oppenheimer on the Szilard petition, Oppenheimer replied:

> ...that he thought it improper for a scientist to use his prestige as a platform for political pronouncements. He conveyed to me in glowing terms the deep concern, thoroughness, and wisdom with which these questions were being handled in Washington. Our fate was in the hands of the best, the most conscientious men of our nation. And they had information which we did not possess. Oppenheimer's words lifted a great weight from my heart. I was happy to accept his word and his authority. I did not circulate Szilard's petition. Today I regret that I did not.(30)

Survey results which reflected a lack of consensus among scientists about the military use of the bomb were published after the war (1948); a questionnaire had been circulated among scientists one month before the bomb was exploded, but its results were kept secret. Although the questionnaire was worded so ambiguously that only those who had a strong pro-use position could be identified clearly, only 15 percent were for immediate military use to end the war quickly. Thus, the public learned belatedly that scientists were human and torn by conflicting values.

In the decade following the war, appalled by deteriorating relations with the Soviet Union and the predicted weapons race, scientists who had opposed the use of the bomb through the wartime Federation of American Scientists. This organization still publishes the Bulletin of the Atomic Scientists, which, since 1947, has clocked the approaching likelihood of nuclear war. The Pugwash conferences, which began in 1957, were an attempt to increase the scientist-to-scientist dialogue between the Allied Nations and the Soviet Union at a time when the government-to-government contact was less than diplomatic.

For many scientists, the most obvious release from their varying degrees of guilt lay in the conversion of nuclear power from military to peaceful uses. Their buoyant optimism was reflected in Eisenhower's "Atoms for Peace" speech, delivered to the United Nations on December 8, 1953; they foresaw the production of vast amounts of cheap energy for a rapidly growing world – the breeder was to be the model – and the development of new techniques, such as radioactive isotopes, for the study and treatment of disease, as well as untold benefits to agriculture and industry. The peaceful harnessing of the atom would help balance the equation of a lifetime's scientific endeavor; the destruction of Hiroshima and Nagasaki would not be the only legacy.

Despite efforts on the part of many scientists to consider human values and planetary survival, their activities and publications were directed primarily to other scientists or to the governments; and neither group was much attuned to the public at that time.

THE SCIENTIFIC COMMUNITY

The years of secrecy, specialized knowledge, and shared experience had forged a new entity – the scientific community. Alumni of the Manhattan Project, The Radiation Laboratory, and related research groups, this new community learned how difficult it was to control the fruits of their labor. It was their first lesson in the limits to science: Scientific knowledge is but one variable in the matrix of social, economic, and political factors within which decisions involving science itself are forged. The broad notion of a scientific community has persisted but, if the scientific "community" exists as such, it is even more heterogeneous today than it was 30 years ago.

What are the characteristics of today's scientific community, and what generates the confidence expressed toward it by the American public? Before World War II it was a small network of underpaid researchers. It grew rapidly between 1956-66; the shift from "Little Science" to "Big Science" was reflected in a 154 percent increase in the number of persons employed as scientists and engineers during this period. Employment in all other sectors rose only by 24 percent.(31)

	Engineers	Scientists	Total
1956	642,400	224,700	867,100
1966	996,000	416,800	1,412,800

As the community grew in size, it grew in diversity and complexity; its members are also diverse and stratified intellectually, emotionally, and politically, as well as by education, achievement, the nature of the work they perform, and the institution within which it is carried out.

The difference between basic and applied science, between theoretical physics and engineering, is often obscured in the public mind. In part, this is because the distinction is not always as clear as basic scientists would like to think it is; but the confusion also results from the fact that the media and opinion polls lump basic and applied science, and scientists and engineers, together. It is not unlikely that, when a Harris pollster asks the mythical average American what he or she thinks of a scientist's ability to solve the problems related to nuclear power, the respondent is thinking Einstein, while the issues to be dealt with – nuclear waste management or improved reactor safety – are less scientific than technical.

This diversity fits the diversity of institutional arrangements within the scientific community. In addition to college and university laboratories, there are defense laboratories, government laboratories, nonprofit laboratories, independent laboratories, mission-related laboratories, and perhaps even a nineteenth century holdover – someone working alone, unaided by federal funds, in his or her cellar laboratory.

This diversity, which has been apparent to scientists and engineers since science was recognized as a separate field of endeavor, is now becoming apparent to the public. This new recognition is producing some bewilderment; scientists, who are supposed to have "the" answer, are disagreeing on ostensibly scientific questions. Increasingly, however, the questions are not scientific or even technical; they go well beyond the bounds of science and technology to involve economics, politics, basic values, and, ultimately, ethics.

The education of scientists and engineers has not prepared the majority of them to answer the kinds of questions that the public is asking. Acquiring highly-technical knowledge means perforce narrowing one's focus in order to master a large body of scientific material. It is often difficult to believe that the answer found through scientifically-valid procedures is not "the" answer.

Clearly, the first task of the scientific community, is to begin to establish for itself criteria of what are facts and what are values. Because the larger community, the public, expects so much of scientists, it has become the responsibility of the scientists to begin to differentiate for the public questions that have more to do with the probabilities, risks, and hard choices.

"Science for science's sake," once an accepted shibboleth of the profession, is now rarely defended. Serious ethical questions concerning the "limits to inquiry"(32) and political questions regarding the regulation of research – how much and by whom(33) – have been raised by scientists.

Although much of the impetus toward greater social responsibility was generated within the scientific community(34), an equally-significant stimulus came from the public, which, in turn, influenced political decisions. As budgets shrank and disillusionment with technology grew, the public and its political representatives demanded more direct participation in the decision-making processes affecting every aspect of science and technology.

Today the challenge for scientists is twofold: There is the need to put the limits to scientific knowledge into perspective within the community, while at the same time interpreting this knowledge to the public.

THE CASE OF NUCLEAR WASTES

The questions about the nuclear fuel cycle far outnumber the solution: Will plutonium recycling lead to nuclear weapons proliferation? Will the potential for plutonium theft increase not only the terrorism but the degree of authoritarianism needed to protect society from the terrorists? How safe are storage mechanisms and sites? Who will determine the choice of these sites? Can the present generation assume responsibility for future generations?

In short, can waste management problems be solved in a manner that satisfies economic, technological, social, and ethical concerns?

The questions have created sharp differences of opinion among scientists and engineers. The charges and countercharges rocketing back and forth between different points of view threaten to diminish the significance of scientific advice. Scientists Rasmussen, Bethe, Wilson, Grey, Ehrlich, and Taylor have been identified with unwavering points of view; non-scientist activists, such as Naderand Lovins have become media performers; lobbying organizations, the Union of Concerned Scientists and the Federation of American Scientists have joined the fray; and journals, Environment and Nuclear News have kept up the momentum. Even the unfortunate vocabulary – hard versus soft,(35) establishment versus radical(36) – bespeaks the polarization.

Increasingly, the rhetoric is that of war. The battle between the Nobelists was triggered in December 1974 when Hans Bethe and nine other Nobelists published a pro-nuclear manifesto: "On any scale the benefits of a clean, inexhaustible fuel far outweigh the possible risks. We can see no reasonable alternative to an increased use of nuclear power to satisfy our energy needs."

The response from the Union of Concerned Scientists was geared to coincide with the thirtieth anniversary of Hiroshima, August 6, 1975. More than a thousand members of the American Technical Community, including eight Nobel prize winners, anti-and go-slow-nuclear-energy individuals, industrialists, and academics, launched their missile to the President and Congress of the U.S.:

> There was once widely-shared enthusiasm among scientists that nuclear fission would represent an inexhaustible new energy source

> for mankind, valuable because it would be safe, inexpensive, and non-polluting. This early optimism has been steadily eroded as the problems of major accidents, long-term radioactive waste disposal, and the special health and national security hazards of plutonium became more fully recognized. It also became clear that the nuclear power proponents failed to appreciate in due course the practical problems that could interfere with the implementation of this new technology, of how companies and individuals might fail to achieve the high level of performance required to safeguard the prodigious quantities of radioactive materials accumulating in a country-wide nuclear power program and thus enhance the risks of serious accidents.(37)

The battle was on.

The intense and heated differences within the scientific community are not difficult to understand; they raise fundamental questions about governance and democracy. With a nuclear-powered economy, there would have to be more governmental control of material and human beings than ever before.

Even if there is a long-term moratorium on new nuclear-energy production, the waste management problem will not go away. By 1985, one year after George Orwell's 1984, some 10,000 tons of commercial spent fuel will have to find a home. Military waste-storage, a factor about which even the well-educated segment of the public knows little, presents an even greater challenge: Although the total radioactivity is comparable to the commercial, it is dispensed in a much larger volume and in a form which will be extremely expensive and difficult to manage.

Economic, legal, and political systems will ultimately determine the outcome of these issues. At present, the scientific community must begin to depolarize the battle by illuminating the nature of its internal disagreements so that the public and politicians can make decisions in less emotionally-loaded circumstances. Not participating in adversary situations where differences are stressed might represent a small step forward; television programs such as the Advocates, or a debate between Rasmussen and Nader force overstatement and certainty where less of both are needed.

It is difficult for scientists and the public to learn to live with doubt and uncertainty. They are, however, the hallmarks of contemporary civilization. There are no technologies that do not involved trade-offs or risks. The public has yet to be educated about these facts. Secrecy has served neither "we" nor "they" well; it has led to doubts and skepticism on everyone's part and has inhibited open debates. It is the honest doubts that have to be explored in public now.

Senator Edmund Muskie said that he was looking for "a one-armed scientist" because every time he asked a scientist a question he received an answer that began, "On the one hand...While on the other..." In the long run, that might turn out to be the best kind of scientist for the scientific community to produce and the public to understand.

NOTES

(1) Dr. John M. Deutch, Statement on Nuclear Waste Management Before the Subcommittee on Oceanography House Committee on Merchant Marine and Fisheries, May 15, 1978.

(2) Rachel Carson, Silent Spring (Houghton Mifflin, 1962).

(3) D.H. Meadows, et al., The Limits to Growth, (New York: New American Library, 1972).

(4) Dorothy Nelkin, "Technology and Public Policy," in Spiegel-Rosing, Ina and Price, Derek de Solla, (Eds.), Science, Technology and Society: A Cross-Disciplinary Perspective, (California: Sage Publications Inc., 1977).

(5) Irvin C. Bupp and Jean-Claude Derian, Light Water: How the Nuclear Dream Dissolved (New York: Basic Books, 1978), p. 135.

(6) David Deese, Nuclear Power and Radioactive Waste, (Massachusetts: Lexington Books, 1978), p. 25.

(7) U.S. Department of Energy Radioactive Waste Management Budgets 1970-79.

(8) David White, "Nuclear Power: A Special New Society Survey," New Society, vol. 39, March 31, 1977, p. 647.

(9) "The Windscale Inquiry," New Scientist, June 23, 1977, p. 694.

(10) Louis Harris and Associates, Inc., A Survey of Public and Leadership Attitudes Toward Nuclear Power Development in the United States, (New York: Ebasco Services Inc., August 1975, p. 36.

(11) Jean-Pierre Olivier, "Radioactive Waste . . . The Problem and Some Possible Solutions," OECD Observer, Sept. 1977, p. 13.

(12) Harvey Brooks, Proceedings of the International Symposium on the Management of Wastes from the LWR Fuel Cycle, Denver, Colorado, July 11-16, 1976, p. 52.

(13) Nucleonics Week, Feb. 16, 1978, p. 5.

(14) Carl Walske, President, Atomic Industrial Forum, Inc., informal communication.

(15) Christopher Hohenemser, Roger Kasperson, and Rober Kates, "The Distrust of Nuclear Power," Science, vol. 196, April 1977, p. 33.

(16) Volkmar Hartje and Meinolf Dierkes, Impact Assessment and Participation Case Studies on Nuclear Power Plant Siting in West Germany, (Berlin: International Institute fur Umwelt und Gesellschaft) 1976.

(17) Directorate of Energy Research, Development and Technology Application, Office of Long-Term Cooperation and Policy Analysis, 1977 IEA Reviews of National Energy Programmes, p. 108.

(18) William Sweet, "The Opposition to Nuclear Power in Europe," Bulletin of the Atomic Scientists, Dec. 1977, p. 41.

(19) Ibid., p. 44, citing Le Monde, Feb. 22, 1977.

(20) BBCI Nuclear Reactor Program, Sept. 21, 1977.

(21) Dorothy Nelkin, Technological Decisions and Democracy, (California: Sage Publications, Inc., 1977).

(22) Patricia Kelly,"Europe's Nuclear Circus," Nature, vol. 271, Feb. 2, 1978, p. 398.

(23) Alice Kimball Smith, A Peril and a Hope, (Cambridge, Mass.: M.I.T. Press, 1971).

(24) Robert, Gillette, "Nuclear Safety," Science, 177, Sept. 1, 8, 15, 1972.

(25) Walter C. Patterson, Nuclear Power, (London: Penguin Books, 1976).

(26) H. Peter Metzger, The Atomic Establishment, (New York: Simon & Schuster, 1972).

(27) Daniel Yankelovitch, et al., Corporate Priorities Program Data, 1977.

(28) Harris, op. cit.

(29) Alvin Weinberg, "The Maturity and Future of Nuclear Energy," American Scientist, vol. 64, Jan.-Feb., 1976, pp. 16-21.

(30) Edward Teller and Allen Brown, The Legacy of Hiroshima, (Garden City, New York: Doubleday & Co., Inc., 1962), pp. 13-14.

(31) Science Indicators , National Science Board , 1975.

(32) Daedalus: Limits of Scientific Inquiry, vol. 107, no. 2, Spring, 1978.

(33) Dorothy Nelkin, "Threats and Promises: Negotiating the Control of Research," Daedalus: Limits of Scientific Inquiry, vol. 107, no. 2, Spring, 1978, pp. 191-209.

(34) American Association for the Advancement of Science, Committee on Scientific Freedom and Responsibility, "Scientific Freedom and Responsibility," (Washington, DC, 1975).

(35) Amory Lovins, Soft Energy Paths: Toward a Durable Peace, (Ballinger, 1977).

(36) Alvin, Weinberg, "Reflections on the Energy Wars," American Scientist, vol. 66, March-April, 1978, pp. 153-158.

(37) "Nuclear Power: A Declaration (To The President and Congress of the United States) by Members of the American Technical Community," Union of Concerned Scientists, Cambridge, Ma., August 6, 1975.

11 The Impact of North-South Politics

Tariq Osman Hyder

An examination of spent fuel management must begin with the assumptions of this volume and their place in the historical and contemporary context of the developed world's nonproliferation strategy. The target area and the validity of these assumptions must be studied in light of the past and evolving positions of the developing countries. The following fundamental assumptions have been made: The spread of commercial nuclear power in the developing countries carries with it the threat of nuclear nonproliferation since spent reactor fuel per se is dangerous in the hands of all but a few responsible states because it could be reprocessed to produce fissile material and nuclear weapons; international management of spent nuclear fuel is necessary because existing international IAEA safeguards are inadequate to detect and, probably, to prevent a state from moving to a weapons capability in a time frame which would allow sanctions and international political action.

If these assumptions represent the bedrock of an existing international consensus on nonproliferation or the fundamental elements on which such consensus is evolving, then it could be said that the backend of the nuclear fuel cycle predominates the nonproliferation issue. Such an approach would lead to an examination of the domestic and international implications of international spent fuel management in terms of siting, regional responses, cost benefit analysis, institutions, and legal problems; and the complex issue of reprocessing could be settled by asserting that this concept would offer one readily-available alternative to storage and/or later reprocessing under international management.

This essay briefly examines the practicality of such an approach. A survey of the trends in the nuclear technology transfers policy of the United States, the prime mover in curbing proliferation, identifies the factors which have helped or hampered nonproliferation.

Phase I was characterized by technological restraints, coupled with the first and only initiative for universal nonproliferation. Three months after the end of World War II, on November 15, 1945, officials of the U.S., British, and Canadian Governments met in Washington and decided to adopt a policy of secrecy in the nuclear field until a system was established for the effective international control of atomic energy. They also decided to buy all available quantities of uranium, thereby preventing proliferation by the denial of the basic source material as well as the technical knowledge.

In December 1945, the USSR accepted a joint U.S.-U.K. proposal to establish an Atomic Energy Commission within the United Nations; this was to consist of representatives from the 11 countries on the Security Council and Canada. On January 24, 1946, the United Nations, at its first session, adopted Resolution 1(1) which set up the Atomic Energy Commission. U.S. policy, which evolved out of the Acheson-Lilienthal Report, was presented by the U.S. delegate Bernard Baruch to the first meeting of the Commission in June 1946. This first and only truly international proposal to control nuclear energy and its attendant potential dangers called for broad international controls:

> ...the creation of an International Atomic Development Authority, to which should be entrusted all phases of the development and use of atomic energy, starting with the raw material and including: 1) Managerial control or ownership of all atomic energy activities potentially dangerous to world security; 2) Power to control, inspect, and license all other atomic activities; 3) The duty of fostering the beneficial uses of atomic energy; (and) 4) Research and development responsibilities of an affirmative character intended to put the Authority in the forefront of atomic knowledge and thus enable it to comprehend, and therefore to detect, misuse of atomic energy.

The U.S. proposal was not acceptable to the USSR, and USSR counter-proposals were not acceptable to the U.S. Inconclusive negotiations ended with the dissolution of the Atomic Energy Agency by General Assembly Resolution 502(VI) on January 11, 1950.

The joint U.S.-U.K.-Canadian policy of technological secrecy did not last after 1949. The discovery of worldwide uranium sources and growing military nuclear programs in the USSR, the U.K., and France led not only to the first spurt of proliferation but eventually to the reversal of U.S. policy on nuclear technology transfers.

Phase 2 began with President Eisenhower's December 8, 1953 speech to the United Nations General Assembly, in which he proposed the creation of the IAEA – established in 1957 – and the Atoms for Peace Program. The U.S. Atoms for Peace Program was criticized for promoting nuclear energy before it was economical. The Program was ineffectual in implementing its goals; sensitive technologies were prematurely declassified, and guarantees of peaceful use were so loosely written as to seem to permit explosives. Critics do hold, however, that it had two major accomplishments: It isolated the commercial fuel cycle from nuclear weapons use; and it established a general climate against the spread of nuclear weapons capabilities.(1)

On the institutional side, the Nonproliferation Treaty was negotiated in 1970, and has now been ratified by over 100 countries.

Since 1973, the complacency of earlier years has been shattered by a number of factors: Energy resources have presented a growing problem, especially after the Arab oil embargo and the rise in oil prices. Uranium prices have risen; and the U.S. has failed to guarantee enrichment contracts. There has been a concommitant and growing perception in Germany, France and Japan that their economic security lay in developing the next stages of nuclear power technology – reprocessing, plutonium recycle and breeder reactors. France and Germany have made a determined push to sell nuclear reactors abroad, and to include enrichment and/or reprocessing plants. There has been a growing public debate within the developed world on the environmental aspects of commercial nuclear power generation. Concern has grown in some developed countries over the potential proliferation of the impending plutonium age. India's nuclear explosion of 1974, the nuclear potential of Israel and South Africa and Taiwan's hot cell for reprocessing – now reportedly dismantled – were seen as threats to world peace. Under U.S. pressure, France cancelled the sale of a reprocessing plant to South Korea. And Canada violated its contract to supply nuclear fuel elements for Pakistan's Canadian-built Kanup reactor in order to pressure Pakistan into cancelling a French reprocessing contract.

Phase 3 is concerned with the evolving U.S. response to these events. It can be studied in two different but complimentary strands, administrative policy and Congressional legislation. In 1974, the U.S. and a group of supplier countries reached an understanding on safeguard requirements for nuclear technology transfers to non-nuclear states that were not parties to the Nonproliferation Treaty. This understanding grew out of the recommendations of the Zanger Committee, and led to the definition of a Trigger List of materials and technology.(2)

In the aftermath of the rather unsatisfactory NPT Review Conference in 1975, seven of the most important suppliers began to meet secretly in London to evolve a common policy. This London Club has played an increasingly important role in determining the current restraints phase of U.S. policy.

The beginning of the London Club, and President Ford's October 28, 1976 policy announcement on nuclear technology transfers represent the two major elements of the Administration policy. As a candidate, Carter made nonproliferation a central tenet of his proposed policy, and he has adhered to it. His declaration of April 7, 1977 follows:

> ...First, we will defer indefinitely the commercial reprocessing and recycling of the plutonium produced in the United States nuclear programs. From our own experience we have concluded that a viable and economic nuclear power program can be sustained without such reprocessing and recycling.
>
> Second, we will restructure the United States breeder reactor program to give greater priority to alternative designs of the

> breeder, and to defer the date when breeder reactors would be out into commercial use.
>
> Third, we will redirect funding of the United States nuclear research and development programs to accelerate our research into alternative nuclear fuel cycles which do not involve direct access to materials useable in nuclear weapons.
>
> Fourth, we will increase U.S. production capacity for enriched uranium to provide adequate and timely supply of nuclear fuels for domestic and foreign needs.
>
> Fifth, we will propose the necessary legislative steps to permit the United States to offer nuclear fuel supply contracts and guarantee delivery of such nuclear fuel to other countries.
>
> Sixth, we will continue to embargo the export of equipment or technology that would permit uranium enrichment and chemical reprocessing.
>
> Seventh, we will continue discussions, with supplying and recipient countries alike, of a wide range of international approaches and frameworks that will permit all nations to achieve their energy objectives while reducing the spread of nuclear explosive capability. Among other things, we will explore the establishment of an international nuclear fuel cycle evaluation program aimed at developing an alternative fuel cycle and a variety of international and U.S. measures to assure access to nuclear fuel supplies and spent fuel storage for nations sharing common nonproliferation objectives.(3)

U.S. policy now rests on a strategy designed to deal with elements of the proliferation problem – the motivation to acquire nuclear weapons, and the technical capability to produce them. In the first case, the strategy is aimed at diminishing the motivations to acquire explosive capability by reducing the role of nuclear weapons in world politics and eventually eliminating them, reducing nuclear weapon force levels, halting nuclear tests, gaining wide adherence to the NPT, and strengthening the IAEA safeguards system.

In the second case, the U.S. hopes to limit technical nuclear capabilities by gaining acceptance for the following measures: The U.S. will offer continued assistance only if full scope safeguards are applied; restraints would be placed on the transfer of sensitive facilities and techniques, especially enrichment and reprocessing; exploring incentives such as assured fuel supplies, spent fuel and waste storage facilities, and assistance in non-nuclear resources would be explored; and there would be an international evaluation of current and future nuclear fuel cycles.

A rather abrasive initial Carter Administration approach did not lead to unity with Japan, Germany, and France; and it brought forth charges of nuclear neo-colonialism and technological hegemony by Third

World countries. Hence, the setting up of INFCE can be seen as a fallback measure. The recently announced joint agreement of the now 15-member London Club on a uniform policy on nuclear technology transfers,(4) and the Nuclear Nonproliferation Act of 1978 (Public Law No. 95-242) represent the apex of present U.S. nonproliferation policy.

A most significant legislative initiative is contained in the Symington Amendment, P.L. 94-329; this amends the Foreign Assistance Act by adding a new section, 699, which specifies that no economic or military assistance be given to countries delivering or receiving nuclear reprocessing or enrichment facilities under multilateral management. The Symington Amendment further states that the recipients of U.S. nuclear fuel and facilities must accept IAEA safeguards. The U.S. Nuclear Nonproliferation Act of 1978 essentially requires that U.S. nuclear cooperation with all countries, irrespective of previous agreements, be conditioned on the recipient country's acceptance of full scope safeguards within two years and a U.S. veto on reprocessing.

The target area for the Northern world's proliferation concern can be broken down into the developing countries, and the so-called "pariah" countries of Israel, South Africa, South Korea, and Taiwan. A glance at the world list of nuclear power plants on Table 11.1 shows that currently there are no operating power reactors in the "pariah" countries and there is a total of five in three non-communist developing countries: Argentina has one; India, three; and Pakistan one.

Atomic Industrial Forum's Industrial 1976 Survey had a most optimistic projection for planned reactors; the general target area covered 18 countries – Israel, South Africa, South Korea, Taiwan, Argentina, Brazil, Egypt, India, Indonesia, Iran, Mexico, Pakistan, Phillipines, Thailand, Yugoslavia, Kuwait, Libya and Rumania.

Despite the assumption that the existing IAEA safeguards regime is inadequate, the status of existing safeguards(5) must be highlighted. Currently 56 states are party to the NPT with NPT safeguards agreements in force and NPT nuclear weapons states; 44 additional states are party to the NPT but do not yet have NPT safeguard agreements in force, although four have IAEA safeguards on all their nuclear activities; 39 non-NPT states have no significant nuclear activity; two non-NPT states have no significant nuclear activity; two non-NPT states are nuclear weapons states as recognized by the NPT; and eight non-NPT states have all their nuclear activity covered under IAEA safeguard agreements. These latter de facto full scope safeguard states are Argentina, Brazil, Chile, Colombia, Indonesia, Pakistan, Portugal, and Turkey.(6)

At present, Egypt, India, Israel, South Africa, and Spain are the only states which have nuclear activities not subject to IAEA safeguards; of these, only Egypt is an NPT signatory. India has two fuel fabrication plants, three reactors, and two reprocessing plants; Israel has a large research reactor and a pilot reprocessing facility; South Africa has a pilot enrichment plant; Egypt has a small research reactor; and Spain has an unsafeguarded power reactor which it operates jointly with France.

Because one of the main assumptions behind this volume is the link

Table 11.1

	Total	Operable	Under Construction	On Order
1. Argentina	2	1	1	
2. Austria	1		1 (complete)	
3. Belgium	7	3	4	
4. Brazil	3		2	1
5. Bulgaria	4	2	2	
6. Canada	24	9	11	4
7. Czechoslovakia	5	1	4	
8. Egypt	1			1
9. Finland	4	1	3	
10. France	49	14	22	12
11. German Democratic Republic	7	3	4	
12. German Federal Republic	29	11	9	7
13. Hungary	4		4	
14. India	8	3	5	
15. Iran	4		2	2
16. Italy	9	4	1	4
17. Japan	28	15	10	1
18. Luxemburg	1		1	
19. Mexico	2	2	2	
20. Netherlands	2	2		
21. Pakistan	1	1		
22. Philippines	2			2
23. Poland	1			1
24. Rumania	1			1
25. South Africa	2			2
26. South Korea	5		3	2
27. Spain	16	4	7	5
28. Sweden	12	7	2	2
29. Switzerland	6	3	2	1
30. Taiwan	6		6	
31. United Kingdom	39	33	6	
32. U.S.A.	203	74	77	52
33. U.S.S.R.	36	22	11	2
34. Yugoslavia	1		1	

Taken from "World List of Nuclear Power Plants," 8/78, pp. 67-85. (30 MWE and over) between commercial nuclear power, spent fuel, and proliferation, it is necessary to examine the history of this apparent link. The following quote is taken from a 1978 statement by the Director General of the IAEA:(7)

> During the period 1945-54, three nations became nuclear weapon states. During the next ten years, 1955-64, a further two nations formed this group. During the next 10 years, 1965-79, one further nation demonstrated that it had the capacity of exploding a nuclear device. Now, let us look at this end, in other words, at this diminishing rate of horizontal proliferation, against the development of peaceful nuclear power. In 1954, the first five-megawatt nuclear-power plant began operation. By 1968, 9,000 megawatts of nuclear power were in operation. By the end of last year, 1977, the figure had risen to 100,000 in 19 countries. The likelihood is that by 1985, peaceful nuclear power capacity will have doubled. . . .I would like to emphasize this point which is the closest possible refutation of the arguments that are frequently advanced that nuclear power inevitably leads to nuclear weapons.

Has any country which has gone nuclear, or is suspected to have a nuclear capability, utilized spent fuel from a commercial reactor, or contravened IAEA safeguards? The U.S., USSR, Great Britain, France, and China made use of what are euphemistically called "dedicated" facilities – gaseous diffusion plants, reactors and reprocessing plants – specifically set up to produce weapons-grade material; India used a research reactor and a reprocessing plant, neither of which was covered by IAEA safeguards; Israel and South Africa have no operating power reactors and rely upon unsafeguarded facilities. Hence, proliferation is possible in terms of reactors alone: Power reactors are the object of international spent fuel management; significant-sized research reactors produce fissile material suitable for Pu239 separation; significant experimental fast reactors utilize strategic amounts of enriched weapons grade U238 or Pu239; and, of course, there are the "dedicated" facilities.

Even without the spread of present enrichment technology and the advent of far cheaper – though as yet experimental – enrichment technologies of centrifuge, jet nozzle, and laser, the spent fuel approach has certain limitations. A country determined to go nuclear could set up a "dedicated facility" for less than the cost of a large commercial reactor, estimated to be about a billion dollars. A recent Congressional study(8) estimated the cost of a modest but adequate natural uranium fueled, graphite moderated reactor at $10.2 million, and that of a complimentary reprocessing plant at around $25 million.(9)

What happens to the 18-country target area if research reactors are introduced? At present, there are only three significant research reactors with nuclear weapons potential: Trombay in India has a 40 MWT; Dimona in Israel, a 26 MWT; and Huactzapu in Taiwan a 40 MWT.

By 1980, taking research reactors(10) alone, the potential target area will have increased from 18 to 23; Chile, Cuba, Thailand, Malaysia, and Venezuela will have joined the group.

Under Article XII A.5 of the IAEA, any country that wants to set up a fast research reactor can claim right to retain – and by implication enrich or reprocess – quantities of fissile material under IAEA safeguards to be used for specified research in reactors. India has operated one fast research reactor, Purina, since 1972, and is constructing a fast breeder test reactor for operation in 1979; and Iraq has signed a letter of intent with France for construction of a PWR plant, to be followed by the eventual construction of a liquid metal fast breeder reactor similar to Phoenix.

Another assumption of this volume is that states could be prevented from acquiring a weapons capability by establishing international management over the spent fuel which might otherwise be reprocessed. This would allow additional time for international political and institutional action to be taken that would decrease risks of weapons proliferation.

To date, no political action has been taken against India and Israel, whose nuclear developments have qualitatively changed the regime envisioned by the NPT. In the case of India, U.S.-Indian nuclear cooperation continues, at least for the next 18 months; Soviet-Indian nuclear cooperation also continues. Although India's neighbor and U.S. ally, Pakistan, was the object of U.S. economic and military sanctions because of its IAEA-approved(11) reprocessing-plant deal with France, no such action has been taken in the case of India. In fact, according to official Indian sources, the U.S. has asked India to join the London Club.

Israel is dependent on the U.S. for its nuclear power program. The mysterious circumstances surrounding the hijacking of a ship full of uranium, two missing lots of weapons-grade fissile uranium 238 from the Apollo plant, and reported U.S. CIA nuclear technology aid all undermine the credibility of the argument to allow time for further international political and institutional action.

The other side of the coin is the evolving response of the target area countries to the developed world's nonproliferation policy. In the context of present political pressure, Israel does not feel obliged to join the NPT regime or its stepchild, the full-scope safeguards regime. South Africa might join, despite recent U.S.-South Africa talks on the issue, but only if the regime feels that international financial and energy sanctions are a credible possibility; severe internal and external unrest might also be a factor in causing it to join. At the present time, South Africa feels that its nuclear potential is an ace card in its relations with emerging and hostile African neighbors; its acceptance of an international safeguard regime hinges on meaningful U.S. action. As client states of the U.S., Taiwan and South Korea are unlikely to challenge the present regime, although South Korea may refuse to join a spent fuel scheme unless sufficient U.S. pressure is applied. NPT negotiations brought out the fundamental objections of India, Brazil, and Argentina to a regime which they characterized as discriminatory. Each of these countries enunciated its sovereign right to produce nuclear explosives for peaceful purposes. However, despite such internal contradictions in the Southern world, it appears that a policy of restraints imposed by the suppliers will lead to an increasing harmony.

The April 1977 Persepolis Conference on the Transfer of Nuclear Technology was summed up by Ambassador Akbar Etemed, President of the AEO of Iran: There were deep concerns held through the Conference regarding the increasing limitations and restrictions being placed on the transfer of nuclear technology. A few even argued that such restrictions are designed to preserve the monopoly of the supplier nations, and to perpetuate the dependency of the recipient countries. Many were concerned about the growing diversion of the policies and practices of the supplier nations from the spirit and the provisions of the NPT. Some considered the recent American policies to abrogate Article 4 of the Nonproliferation Treaty, and to seriously damage the institution of IAEA and its mandate.

At the IAEA Board of Governors Conference, as well as the IAEA General Conference in Vienna (at the General Conference in particular), it was evident that there was Southern dissatisfaction with policies of restraints on transfers of nuclear technology, technical assistance, nuclear fuel, and the continuing concentration of the Agency on safeguards to the detriment of its technical assistance program.

Particular attention was given to paragraphs 5 and 8 of INFCE's organizational session's Final Communique. These stress the need for addressing vertical proliferation and insist that INFCE would not affect on-going national nuclear power developments and existing international agreements on transfers of nuclear technology.

The 32nd U.N. General Assembly included two resolutions on transfers of nuclear technology; both were adopted in November and December 1977.

Is the issue of international management of spent nuclear fuel an isolated technical and institutional phenomenon? What is the alternative? On what assumptions can a nonproliferation regime rest and how specifically can spent fuel management be brought into this wider conceptual context?

In the best of all possible worlds, the attainment of international peace and security, coupled with universal nuclear disarmament, would remove the danger of proliferation which now constitutes the primary element of the arms race. Within the parameters of the possible, the main factors that mitigate against nonproliferation can be outlined: Proliferation is a causal chain going back to the very first explosion; once proliferation has left the domain of the superpowers it becomes a regional problem. The fundamental problem of nonproliferation and, by extension, of any facet of international management of the nuclear fuel cycle has always been the issue of discrimination in terms of military and economic security. The NPT regime has been qualitatively changed by the emergence of Indian, Israeli, and South African nuclear capabilities. The current emphasis on further proliferation and on the selective implementation of safeguards and controls, rather than on the three nuclear capable non-NPT states, will only lead other countries to make decisions that keep their options open. Technical restraints can only delay for a short time the acquisition of a capability status by other states. INFCE is valuable as a technical exercise, but, unless accompanied by political measures, it will fail to prevent plutonium proliferation.

A new international consensus is needed. Both nuclear and non-nuclear countries have to work together to reach it. The main elements of such a consensus, from the perspective of the Southern world, must be:

1) There should be credible evidence of the willingness of nuclear powers to give up nuclear weapons as a military option; stockpiles and threats should be eliminated.

2) All states should be encouraged to commit themselves to the goal of nonproliferation through adherence to the NPT, nuclear-free zones, or IAEA safeguards.

3) All nuclear facilities which are not presently subject to IAEA safeguards should be brought under the IAEA system; all accumulated fissile material must be accounted for under this safeguard system.

4) Agreements should be reached on adequate standards for the physical protection of nuclear material.

5) International cooperation for the development of peaceful nuclear technology in the LDC's must be accelerated; and all joint projects, such as INFCE, should be undertaken in such a way as not to prejudice national nuclear energy programs and priorities or existing international agreements.

The international management of spent fuel would have to fall within these parameters. An international spent fuel regime would have to, in a nondiscriminatory manner, safeguard the economic security of all states, and should include the developed countries as well as the undeveloped ones.

Because the developed countries, with the exception of the U.S., seem to be headed toward the plutonium economy, international cooperation in spent fuel storage must be part of the broader international nuclear regime, and must include fuel supplies.

This leads to the question of assured fuel supplies and reprocessing, or co-processing, on demand. Domination of this area by the developed countries would not be acceptable to the developing countries. The search for closed fuel cycles can, in the economic context, only be broken by a nondiscriminatory international regime. Fuel supplies, spent fuel storage areas, and potential reprocessing facilities would have to be co-located in developing and developed countries alike. The actual structure of international institutions to deal with these matters is secondary in importance to their being controlled in a nondiscriminatory manner.

Hypothetically this approach could win acceptance in the Southern world. The argument can be made that the study of this particular topic cannot be linked to the overall problem. However, a realistic assessment of the probabilities of success for any proposal, however limited it may be in scope, must consider the mix of factors involved and weigh the relative importance of each factor.

NOTES

(1) Joseph S. Nye. Deputy to the Assistant Secretary of State for Technical Assistance, Science and Technology. Address of June 30, 1977 to the Houston Rotary Clubs. See also his article: "Nonproliferation, a Long Term Strategy" in Foreign Affairs, April, 1978.

(2) INFCIRC 209 act Add. 1-8. (IAEA document) For an extended discussion of this issue, refer to the UN Disarmament Yearbook. Vol. l: 1976. Chapter IX.

(3) For the most recent elaboration, see Ambassador Young's statement on Peaceful Nuclear Cooperation to the UN Ad Hoc Committee of the UN Special Session on Disarmament on June 9, 1978, Press Release USUN-59(78), June 9, 1978, also UN doc A 5-10 AC 1 PV.5.

(4) Nuclear Suppliers Group. Guidelines for Nuclear Transfers. See INFICRC/254, Feb. 1978 (IAEA document).

(5) Based on Article on IAEA safeguards in IAEA Bulletin. Vol. 19, no. 5, Oct. 1977.

(6) Turkey is now a signatory to the NPT. For its timetable on ratification, see Prime Minister Ecevit's statement to the UNSSUD on June 1978 (A/S/-10/PU 15. UN document).

(7) Director General of the IAEA, Dr. Eklund's statement in UN document A/S/10/PV 13.

(8) Nuclear Proliferation Fact Book, September 23, 1977, printed for the use of Committee on International Relations, U.S. House of Representatives, and the Committee on Government Affairs, US Senate, U.S. Government Printing Office, no. 152 070 04290-1.

(9) If built in the U.S. Concrete has been estimated to be the most expensive single item for plant construction; probably lower concrete costs as well as lower labor costs would further depress the cost for a developing country.

(10) SIPRI's 1971 Yearbook (Section Chart on Survey of Research and Experimental reactors).

(11) The Pakistan French reprocessing plant agreement (INFICRC 239 IAEA document which was covered by a 20-year pursuit technology safeguard was approved by the IAEA's Board of Governors which included the assenting vote of the U.S. representative.

Conclusion

The fundamental working assumption of this volume is that nations only engage in serious international cooperation for problems which exceed their individual capacities. Although the concept of international management of spent nuclear fuel may still be identified primarily with the U.S., countries without near term plans for local reprocessing might be interested; such would include Canada, Sweden, Austria, South Korea, Taiwan, and the Philippines, and perhaps even some countries planning to reprocess commercially.

What incentives could there be for nations to participate in some form of international management for spent fuel? Not one country yet has a fully operational closed nuclear fuel cycle. And the prospects for any, with the exception of the five or six most technologically-advanced, obtaining a full fuel cycle for the next 10 to 20 years still seem quite remote.

Even the technically-simple operation of spent fuel storage has, for many reasons, such as delays elsewhere in the fuel cycle and political or legal implications, become a severe problem in many countries. India was the first of the less-developed countries to have operation of a power reactor (Tarapur) seriously threatened by the spent fuel storage problem, but it is unlikely to be the last. Opposition to nuclear power programs in general and to specific sites for reactors or other fuel cycle facilities has increased dramatically in the mid-and late-1970s. Countries such as Denmark and Thailand have rejected, at least for now, the initiation of nuclear power programs on environmental grounds. Environmental opposition has, in part, been responsible for a two year delay in Turkish plans for a nuclear power plant on its Mediterranean coast.(1)

The first power reactor in the Philippines' nuclear program remained under environmental attack as of late 1978 as a result of the questionable seismic characteristics of its site. In addition, it is increasingly the target of international protests against nuclear exports to developing countries by anti-nuclear groups in Australia, Canada, Europe, Japan, and the U.S.(2)

Brazilian and Iranian nuclear power programs face an increasing potential for environmental opposition after experiencing corruption and other financial problems, such as huge cost increases, severe technical problems (Brazil), and deep social unrest.(3)

The specific question of ultimate disposition for spent fuel, which may not be reprocessed in some cases, and for high level and other wastes from reprocessing also will be of increasing concern to the less-developed countries. The Austrian rejection of a completely-built nuclear power plant stands as a stark example of the political importance of the spent fuel disposition question.(4) Canada, Switzerland, the United Kingdom, and the U.S. face severe and frequently overwhelming local and national environmental opposition even to the taking of core samples for radioactive waste disposal tests in promising geological formations. Countries will also continue to be concerned about already-established radioactive waste disposal sites; for example, Kentucky closed a low level waste burial site after a long controversy over seepage and ground water contamination.(5)

Finally, the option of reprocessing and waste disposal does not simplify the final disposition problem. Many technical studies, papers, and debates have led to a growing consensus that there is no clear waste management advantage to either reprocessing or not reprocessing. Reprocessing itself remains a complex and costly operation: the very expensive contracts let by Cogema of France and British Nuclear Fuels Limited in 1978 could not be fulfilled; serious technical problems shut down the Tokai Mura reprocessing plant in Japan(6) and the BNFL reprocessing plant at Windscale;(7) and anti-nuclear groups in France have been joined by local political leaders and union representatives in their continuing protest against planned expansions at the Cap de la Hague reprocessing plant.(8)

This brief sketch of problems confronting the nuclear fuel cycle is intended to demonstrate that spent fuel management requires international cooperation. International spent fuel management offers a wider range of alternatives to countries with, or planning for, nuclear power programs. It offers an alternative to deciding very quickly to proceed with what is certainly a complex and expensive operation – commercial reprocessing of spent light water reactor fuel. Its intent is not to delay or prevent nuclear power programs; on the contrary, some form of international spent fuel management might well be essential to their survival.

Neither is spent fuel storage posited as a general solution to nuclear proliferation problems. It constitutes only one part of a complex set of political and institutional means to establish adequate national and international security. In a number of specific circumstances, it could help provide additional time to address the threatening international security problems posed by reprocessing and plutonium management.(9)

Despite the wealth of forward-looking and constructive analyses contained between the covers of this book, the concluding paper must add a pessimistic note: The likelihood of rational economic and political analysis prevailing as the dominant force in nuclear energy policy in the near future is not great. Countries seem to set the courses of their nuclear-fuel-cycle programs more by idle hopes and grand dreams than by calculations of economic and political advantage.

This volume reviews the technical and economic credibility of internationally-managed spent fuel storage as a nuclear-fuel-cycle option. It also opens a number of political and institutional possibilities for increased international oversight of spent fuel. Definitive design work cannot, of course, be done in an academic setting; engineers, accountants, and economists must do the future detailed work on which actual programs stand or fall. Nevertheless, certain basic propositions do emerge from the foregoing chapters.

In Chapter 7, Miller showed that interim storage in away-from-reactor pools is an adaptation of existing technology for at-reactor storage. There are differences of opinion about how long it is prudent to store fuel in this mode, but no one doubts that it can be done for at least 20 years. Miller concludes that this form of storage may be safe for many decades, and, in any event, 40 year storage is not at all unreasonable to anticipate.

A number of new technologies are in varying stages of development for dry storage, but none of these represents more than a modest extension of existing engineering knowledge and principles. Canadian work on dry storage is furthest advanced. In part, this is because it became apparent that CANDU spent fuel is uneconomic to reprocess in today's uranium market. The desire to hedge against future reverses led Canada to embark on a program for retrievable surface storage in a passive dry mode, a compromise between the immediate retrievability of wet storage and the security of permanent disposal. Certain major modifications would be necessary to adapt this technology to light-water-reactor spent fuel because size, heat rejection loads, and other factors differ, but the engineering and testing required is simple and straightforward.

Miller also points out that the spent fuel's radiation barrier offers a useful – if not perfect – nonproliferation barrier for long periods of time. Doubtless over 40 years the strength of this barrier is much reduced when the fuel exits the reactor; nevertheless, it is still a formidable obstacle to illlicit reprocessing.

Greer and Dalzell indicate that, under a wide range of circumstances, centralized-spent-fuel storage should be economically attractive compared to new at-reactor basins. Countries with modest or small nuclear programs could generate significant economies of scale in storage costs through participation in a regional or multinational spent-fuel-storage regime. In many cases these economies would more than offset the substantial one-way transportation penalties inherent in a regional away-from-reactor scheme. For nations with larger annual outputs of spent fuel, the advantage of regional arrangements is less clear. Although the local transportation costs would be smaller, the advantage of building larger away-from-reactor units is swamped by the economic case for buiding additional capacity at reactors. Thus, from the economic point of view, participation in some mode of centralized storage is persuasive.

Although the subject of reprocessing is, in general, beyond the purview of this book, recent work indicates that the net benefit of plutonium is enhanced if it is used at least fifteen years hence rather than immediately. A recent Rand Corporation Study(10) computed the present discounted future value of spent fuel, taking into account the streams of net costs and net benefits which should accrue at various times in the future. Whether for light-water recycle or breeder use, extracting the plutonium and reusable uranium in spent fuel does not, the study implied, make economic sense before at least 1993 in any but the most unlikely sorts of scenarios. At that, this study is biased toward reprocessing because it accounts for the cost of disposal in the once-through fuel cycle option, but not in any of the reprocessing options.

Moreover, Chapter 1 suggests that no insuperable legal, institutional, or geopolitical hurdles stand in the way of multinational cooperation for construction of a storage regime. Features of a desirable institutional structure are relatively easy to discern; they involve some commitment to furthering nonproliferation objectives, but, most of all, they avoid threats to any country's fuel cycle. Institutional models for the kind of internal structure which might be desirable abound: Paradigms range fromthe Eurochemic Charter to the agreements between France and Switzerland for the operation of the Basel-Mulhouse International Airport. Some experience with past international enterprises suggests that over-development of the institutional structure may be fruitless. As with domestic endeavors, new enterprises learn best by experience.

The relative simplicity of institutional demands for international management of spent fuel is exceedingly important. No new international structures would be required for returning spent fuel to supplier nations, or for increasing passive international surveillance of national storage facilities; and these two alternatives are, in and of themselves, very important possibilities.

Very little new structure would be required for long term storage at existing reprocessing plants or at a national facility, such as Barnwell, which could be given some degree of international management. Finally, compared for the obstacles facing the creation of multinational centers for reprocessing, those confronting even new international spent fuel storage facilities would not be overwhelming. The leading problem is finding a site acceptable both to a host nation and to potential participant nations.

Furthermore, the geopolitical criteria developed in Chapter 1 suggest that it would be difficult, but in the end possible, to find an acceptable location for regional storage regimes. Although odious dictatorships should certainly be avoided in the search for a host country, a number of countries exist which could minimally satisfy all of the siting criteria. Motivating an acceptable host to receive foreign spent fuel into its territory might be difficult, but financial, political, and security incentives could offer strong motivation to host countries in a number of cases.

Identifying potential depositors is not an insurmountable problem either. For a number of countries, even in Europe, the prospect of a meaningful breeder program is so remote that no real incentive to reprocess exists other than the need to "do something" about nuclear waste management. Getting nuclear waste and spent fuel out of public sight and out of the country could prove to be a political boon.

Yet major political obstacles to this kind of nuclear cooperation remain. In Chapter 3 Marwah concludes that none of the three major Indian Ocean states – India, Iran, and Pakistan – would be acceptable to both of the others as a host for spent fuel storage; only a distant island site in the remote South Indian Ocean would be satisfactory to all three. In Chapter 5, Broinowski suggests that Japan, a major littoral state, would not encourage spent fuel storage for fear its breeder program might be undermined. In Chapter 6, Gallucci reports that this is also so in Europe, where it is felt that providing the small states with out-of-country storage might somehow injure the commercial energy aspirations of France, Germany, and the U.K. In Latin America, say Johnson and Astiz in Chapter 4, spent fuel cooperation is possible only if the location is in territory deemed neutral by Argentina and Brazil, for example Paraguay and Uruguay; and, even then, it is only remotely possible because of the already far gone nuclear race between the two dominant powers.

It is frequently observed that the Soviet nuclear export system could serve as a model for Western management of spent fuel. Certainly, as Nathanson indicates in Chapter 2, there is no need for a spent fuel policy in eastern Europe because the USSR requires all customers to return all spent fuel, a neat way out of the reprocessing-nonproliferation dilemma. Imitating the Soviets would suggest switching to a strict leasing arrangement for all Western suppliers. Even ignoring the extent to which past exports of fuel as outright sales have preempted the opportunity to pursue this institutional mode, past exports of nuclear technology and the resurgence of European and Japanese technological power have created a polycentric nuclear export market which can no longer be bridled to such an extent. Furthermore, the Soviet Union, acculturated as it has become over the centuries to centralized decision making, can tolerate this sort of monolithic system well, but it is unlikely that such a system could flourish in the loose, decentralized Western economy. And Nathanson suggests that, as Soviet commercial nuclear sales spread to a wider circle of customers beyond the immediate reach of the Red Army, its ability to demand and enforce the kind of discipline it imposes on Eastern Europe will perforce weaken.

A softer system than the Soviet model would have to be devised for the Western nuclear economy. Nevertheless, there is no shortage of appropriate institutions one could superimpose on the nuclear energy system. Creating models is not the immediate problem. The real unanswered question is whether the centrifugal forces of international politics can be overcome by the rather weak attractions of a marginal economic opportunity. It must be said in candor that the prospects are not bright.

The broad factors driving the outward drift are the reinforced environmental and anti-nuclear forces in the developed countries, discrimination concerns in the developing countries, and the push for energy independence everywhere. As Zinberg points out in Chapter 10, papering over environmental and anti-nuclear concerns only strengthens deeply held convictions and stiffens anti-nuclear resolve. Yet much of the European commitment to immediate reprocessing and high level waste disposal is aimed at undercutting public opposition. Reopening the waste disposal debate only creates impediments for European nuclear expansion.

Among the nations loosely-classified as the "group of 77," the catchword "discrimination" has immediate currency. As Hyder points out in Chapter 11, developing nations assert that the bargain represented by the Nonproliferation Treaty demands two responses from the developed nations: First, real reductions in existing nuclear armaments must be made; and, second, nuclear energy technology must be fully-shared, regardless of how necessary on the one hand or risk prone on the other.

Finally, as Poneman points out in Chapter 9, some measure of energy independence might be attained through independence in the nuclear fuel cycle. In the developed and developing countries alike, the theory is that plutonium recycle and the breeder reactor can free nations from resource constraints imposed by others.

One is forced to the conclusion, however, that to a large extent these political arguments are excuses for, rather than causes of, the policies pursued. In Western democracies, there is simply no valid reason to use subterfuge and dissembling in order to cut off public debate on a matter of vital concern – the safety and environmental soundness of methods of managing nuclear wastes. Furthermore, real energy independence, through the breeder or any other nuclear energy policy, is virtually unattainable for any but the most advanced nations. Although it is true that some measure of flexibility and diversity can be introduced in the short run by reliance on nuclear generation, in the long run most countries simply exchange one form of dependency for another. One suspects that what may lie behind some of the "group of 77" arguments is a perception that keeping open a nuclear weapons option, however narrow, has significant political utility.(11)

Thus if nuclear fuel cycle decisions were made on a rational economic basis, centralized-spent-fuel storage would be attractive. On a regional basis, an international storage authority is potentially feasible politically in one or two regions. Nevertheless, broad political tides flow against this possibility and will continue to do so unless something catalyzes renewed progress.

There are, however, other less demanding possibilities for international management of spent fuel. Their nature can be only a matter of conjecture at the present time. Perhaps the ongoing International Nuclear Fuel Cycle Evaluation will lead to a set of findings and points of consensus which will develop support for a more moderate approach to new technology. Perhaps some new initiative by the U.S. or other major fuel exporters will definitively demonstrate the advantages of a cautious approach. Or perhaps it will take a new proliferation event to mobilize the world community. One can only hope that this level of pessimism is unwarranted.

NOTES

(1) Nucleonics Week , November 2, 1978, p. 8.

(2) Nucleonics Week , October 26, 1978, p. 6.

(3) See, for example, Nucleonics Week , Nov. 2, pp. 6, 11, 12, and Oct. 26, p. 13.

(4) See, for example, Energy Daily, Nov. 9, 1978, p. 3.

(5) Nucleonics Week, Oct. 26, 1978, p. 7.

(6) Nucleonics Week, Oct. 12, 1978, p. 8.

(7) Nucleonics Week, Nov. 9, 1978, p. 11.

(8) Nucleonics Week, Oct. 26, 1978, pp. 8-9; and C.F.D.T., National Secretariat for Economy, Employment and Professional Training, "Energy Policy: The CFDT Positions," May 1978.

(9) The recently promoted idea that spent fuel storage creates plutonium mines around the world is seriously in error. Spent fuel storage not only puts off the politically- and institutionally difficult nonproliferation problems posed by reprocessing and plutonium management, it also maintains a much greater radiation barrier to tampering than that offered by any of the co-processing/CIVEX possibilities. (See Bernard J. Snyder, "Non-Proliferation Characteristics of Radioactive Fuel," September 1978.)

(10) Kenneth A. Solomon, "Nuclear Reactor Spent Fuel Valuation: Procedure, Applications, and Analysis," Rand Corporation Report R-2239-DOE (February 1978).

(11) Richard K. Betts, "Regional Insecurities and Nuclear Instability: India, Pakistan, Iran," International Studies Association Convention, February 1978.

Index

About the Contributors

FREDERICK C. WILLIAMS (J.D. -University of Michigan Law School and Ph.D. – University of North Carolina) is actively involved in issues of arms control and security as an attorney for the U.S. Arms Control and Disarmament Agency. During the period 1977-78 he managed the Nuclear Nonproliferation Working Group at Harvard's Center for Science and International Affairs.

DAVID A. DEESE (M.A., M.A.L.D. and Ph.D. - The Fletcher School of Law and Diplomacy at Tufts University) is a research fellow in the Center for Science and International Affairs. He is the author of the book Nuclear Power and Radioactive Waste, and author of articles on the response of international politics and institutions to problems posed by science and technology.

MELVYN B. NATHANSON, University of Southern Illinois.

ONKAR MARWAH, Harvard University.

VICTORIA JOHNSON, Northwestern University.

CARLOS ASTIZ, SUNY at Albany.

RICHARD BROINOWSKI, Department of Foreign Affairs, Australia.

ROBERT GALLUCCI, U.S. Department of State.

MARVIN MILLER, M.I.T.

BOYCE GREER, Donovan, Hamester and Rattien.

MARK DALZELL, Harvard University.

DANIEL PONEMAN, Oxford University.

DOROTHY ZINBERG, Harvard University.

TARIQ OSMAN HYDER, Ministry of Foreign Affairs, Pakistan.

Pergamon Policy Studies

No. 1 Laszlo—*The Objectives of the New International Economic Order*
No. 2 Link/Feld—*The New Nationalism*
No. 3 Ways—*The Future of Business*
No. 4 Davis—*Managing and Organizing Multinational Corporations*
No. 5 Volgyes—*The Peasantry of Eastern Europe, Volume One*
No. 6 Volgyes—*The Peasantry of Eastern Europe, Volume Two*
No. 7 Hahn/Pfaltzgraff—*The Atlantic Community in Crisis*
No. 8 Renninger—*Multinational Cooperation for Development in West Africa*
No. 9 Stepanek—*Bangladesh—Equitable Growth?*
No. 10 Foreign Affairs—*America and the World 1978*
No. 11 Goodman/Love—*Management of Development Projects*
No. 12 Weinstein—*Bureaucratic Opposition*
No. 13 DeVolpi—*Proliferation, Plutonium and Policy*
No. 14 Francisco/Laird/Laird—*The Political Economy of Collectivized Agriculture*
No. 15 Godet—*The Crisis in Forecasting and the Emergence of the "Prospective" Approach*
No. 16 Golany—*Arid Zone Settlement Planning*
No. 17 Perry/Kraemer—*Technological Innovation in American Local Governments*
No. 18 Carman—*Obstacles to Mineral Development*
No. 19 Demir—*Arab Development Funds in the Middle East*
No. 20 Kahan/Ruble—*Industrial Labor in the U.S.S.R.*
No. 21 Meagher—*An International Redistribution of Wealth and Power*
No. 22 Thomas/Wionczek—*Integration of Science and Technology With Development*
No. 23 Mushkin/Dunlop—*Health: What Is It Worth?*
No. 24 Abouchar—*Economic Evaluation of Soviet Socialism*
No. 25 Amos—*Arab-Israeli Military/Political Relations*
No. 26 Geismar/Geismar—*Families in an Urban Mold*
No. 27 Leitenberg/Sheffer—*Great Power Intervention in the Middle East*
No. 28 O'Brien/Marcus—*Crime and Justice in America*
No. 29 Gartner—*Consumer Education in the Human Services*
No. 30 Diwan/Livingston—*Alternative Development Strategies and Appropriate Technology*
No. 31 Freedman—*World Politics and the Arab-Israeli Conflict*
No. 32 Williams/Deese—*Nuclear Nonproliferation*
No. 33 Close—*Europe Without Defense?*
No. 34 Brown—*Disaster Preparedness*

No. 35 Grieves—*Transnationalism in World Politics and Business*
No. 36 Franko/Seiber—*Developing Country Debt*
No. 37 Dismukes—*Soviet Naval Diplomacy*
No. 38 Morgan—*Science and Technology for Development*
No. 39 Chou/Harmon—*Critical Food Issues of the Eighties*
No. 40 Hall—*Ethnic Autonomy—Comparative Dynamics*
No. 41 Savitch—*Urban Policy and the Exterior City*
No. 42 Morris—*Measuring the Condition of the World's Poor*
No. 43 Katsenelinboigen—*Soviet Economic Thought and Political Power in the U.S.S.R*
No. 44 McCagg/Silver—*Soviet Asian Ethnic Frontiers*
No. 45 Carter/Hill—*The Criminal's Image of the City*
No. 46 Fallenbuchl/McMillan—*Partners in East-West Economic Relations*
No. 47 Liebling—*U.S. Corporate Profitability*
No. 48 Volgyes/Lonsdale—*Process of Rural Transformation*
No. 49 Ra'anan—*Ethnic Resurgence in Modern Democratic States*
No. 50 Hill/Utterback—*Technological Innovation for a Dynamic Economy*
No. 51 Laszlo/Kurtzman—*The United States, Canada and the New International Economic Order*
No. 52 Blazynski—*Flashpoint Poland*
No. 53 Constans—*Marine Sources of Energy*
No. 54 Lozoya/Estevez/Green—*Alternative Views of the New International Economic Order*
No. 55 Taylor/Yokell—*Yellowcake*
No. 56 Feld—*Multinational Enterprises and U.N. Politics*
No. 57 Fritz—*Combatting Nutritional Blindness in Children*
No. 58 Starr/Ritterbush—*Science, Technology and the Human Prospect*
No. 59 Douglass—*Soviet Military Strategy in Europe*
No. 60 Graham/Jordon—*The International Civil Service*
No. 61 Menon—*Bridges Across the South*
No. 62 Avery/Lonsdale/Volgyes—*Rural Change and Public Policy*
No. 63 Foster—*Comparative Public Policy and Citizen Participation*
No. 64 Laszlo/Kurtzman—*Eastern Europe and the New International Economic Order*
No. 65 United Nations Centre for Natural Resources, Energy and Transport—*State Petroleum Enterprises in Developing Countries*